Foundation Systems for High-Rise Structures

Foundation Systems for High-Rise Structures

Rolf Katzenbach
Technische Universitaet Darmstadt, Germany

Steffen Leppla
Technische Universitaet Darmstadt, Germany

Deepankar Choudhury
Indian Institute of Technology Bombay, Mumbai,
Maharashtra, India

CRC Press
Taylor & Francis Group
Boca Raton London New York

CRC Press is an imprint of the
Taylor & Francis Group, an **informa** business

CRC Press
Taylor & Francis Group
6000 Broken Sound Parkway NW, Suite 300
Boca Raton, FL 33487-2742

First issued in paperback 2018

© 2017 by Taylor and Francis Group, LLC
CRC Press is an imprint of Taylor & Francis Group, an Informa business

No claim to original U.S. Government works

ISBN-13: 978-1-4987-4477-5 (hbk)
ISBN-13: 978-0-367-13904-9 (pbk)

Library of Congress Cataloging-in-Publication Data

Names: Katzenbach, Rolf, author. | Leppla, Steffen, author. | Choudhury, Deepankar, author.
Title: Foundation systems for high-rise structures / Rolf Katzenbach, Steffen Leppla, and Deepankar Choudhury.
Description: Boca Raton : CRC Press CRC Press is an imprint of the Taylor & Francis Group, an Informa Business, [2017] | Includes bibliographical references and index.
Identifiers: LCCN 2016012107 | ISBN 9781498744775
Subjects: LCSH: Tall buildings--Foundations.
Classification: LCC TH5201 .K38 2017 | DDC 624.1/5--dc23
LC record available at https://lccn.loc.gov/2016012107

Visit the Taylor & Francis Web site at
http://www.taylorandfrancis.com

and the CRC Press Web site at
http://www.crcpress.com

Contents

Preface

Various urban areas in the world are experiencing scarcity of land, and the spatial expansion of buildings and structures is becoming increasingly problematic. High-rise structures are the only solution to this problem. The design, construction, and performance of such high-rise structures mostly depend on the stability of the foundation systems. High-rise structures, such as the Burj Khalifa building in Dubai or the proposed Kingdom Tower in Jeddah, depend upon the performance of their foundation systems. This book is the first to assemble the latest research on the analysis, design, and construction of such foundation systems for high-rise structures.

Based on the authors' own scientific research and extensive experience, and those of researchers from engineering practices, *Foundation Systems for High-Rise Structures* presents the theoretical basics of the analysis and design of all types of foundation systems and explains their application in completed construction projects.

This book deals with the geotechnical analysis and design of all types of foundation systems for high-rise buildings and other complex structures with a distinctive soil–structure interaction. The basics of the analysis of stability and serviceability, necessary soil investigations, important technical regulations, and quality and safety assurance are explained, and possibilities for optimized foundation systems are given. Additionally, special aspects of foundation systems, such as geothermally activated foundation systems and the reuse of existing foundations, are described and illustrated. To complete this book, a comprehensive chapter on the analysis and design of foundation systems and the dynamic behavior of foundation systems for high-rise structures has also been included.

At the end of each chapter, the reader finds an overview of the references used, which is helpful for finding additional information in high-quality literature. To understand the boundary conditions for analysis and design of foundation systems, the standards and regulations are named as well. Due to the complexity of the analysis, design, and construction of the combined pile-raft foundation (CPRF), international guidelines on CPRFs by the International Society for Soil Mechanics and Geotechnical Engineering (ISSMGE) are also included in the Appendix.

The authors thank Mr. Ashutosh Kumar of IIT Bombay for helping to assemble the contents of the chapter about the dynamic behavior of foundation systems. The authors also thank CRC Press/Taylor & Francis Group for publishing this book for professionals in engineering practice and for students and faculty members who will be working in the future in this special field of application.

Rolf Katzenbach
Steffen Leppla
Deepankar Choudhury
Germany & India

Authors

Deepankar Choudhury is a professor at the Department of Civil Engineering at the Indian Institute of Technology Bombay, Mumbai, India, and adjunct professor at the Academy of Scientific and Innovative Research, New Delhi, India. He is an Alexander von Humboldt fellow of Germany, JSPS fellow of Japan, TWAS-VS fellow of Italy, BOYSCAST fellow of India, fellow of the Indian Geotechnical Society, and fellow of the Indian Society for Earthquake Technology. He is an internationally known academic and researcher with expertise in geotechnical earthquake engineering, foundation engineering, computational geomechanics, and dynamic soil–structure interaction. He serves as editorial board member of various reputable journals, including *ASCE International Journal of Geomechanics, Canadian Geotechnical Journal, Indian Geotechnical Journal,* and *INAE Letters.* He is secretary of the technical committee TC 207 – Soil–Structure Interaction and Retaining Walls, former secretary of TC 212 – Deep Foundations, a member of TC 203 of the International Society for Soil Mechanics and Geotechnical Engineering (ISSMGE), and a member of International Building Code (IBC) 1803 on Foundations of USA. A globetrotter, he has given several keynote, plenary, and invited lectures across the world, published several papers in reputable journals, supervised many doctoral and masters students, and been involved in various national and international projects of importance and practical significance. More details about him are available at http://www.civil.iitb.ac.in/~dc/.

Rolf Katzenbach is the director of the Institute and Laboratory of Geotechnics at the Technische Universität Darmstadt, Germany. He is a board member of several international and national organizations. He is a member of the chamber of engineers, a publicly certified expert of geotechnics, and an independent checking engineer providing expertise for national and international courts of justice, arbitration committees, insurance companies, state ministries, building authorities, and large national and international financial institutions and investors. He is responsible for the successful application of the Combined Pile-Raft Foundation at important projects all over the world, and he is a respected specialist for retaining systems, slope stability, and underground constructions, including tunnels for metro systems and high-speed railway lines.

Steffen Leppla is a scientific research assistant at the Institute and Laboratory of Geotechnics at the Technische Universität Darmstadt, Germany. He has worked on several national and international major geotechnical engineering projects concerning high-rise buildings, tunneling, and large mine heaps. His research topics are soil–structure interaction, anchor systems, and salt mechanics as well as the construction and engineering inspection of tunnels. Since 2013, he has been a certified, independent expert and proof engineer for the geotechnics of underground metro systems and tramways. In addition to his research activity at TU Darmstadt, he is currently a visiting professor at St. Petersburg Polytechnic University in Russia.

Chapter 1

Introduction

According to the technical regulations, all types of foundation systems for high-rise buildings and civil engineering structures have to be analyzed for stability and serviceability. Analyses are based on the mechanical parameters of the soil and on the modeling of the soil–structure interaction. The analysis of the stability and the serviceability of the following foundation systems will be explained:

- Spread foundations, for example, strip foundations and raft foundation
- Deep foundations, for example, pile foundations
- Combined foundations, for example, Combined Pile-Raft Foundations (CPRF)
- Special foundations, for example, caisson foundations and well foundations

The development of foundation systems related to the height of the superstructure can be seen by the example of Frankfurt am Main, Germany, where in the last decades several high-rise buildings were founded in the settlement-active Frankfurt Clay (Figure 1.1).

This book covers the basics of stability and serviceability and the necessary soil investigation parameters, the valid technical regulations, and the measures to guarantee the four-eye principle. In addition, special aspects of foundation systems such as geothermal use and the reuse of existing foundations are described.

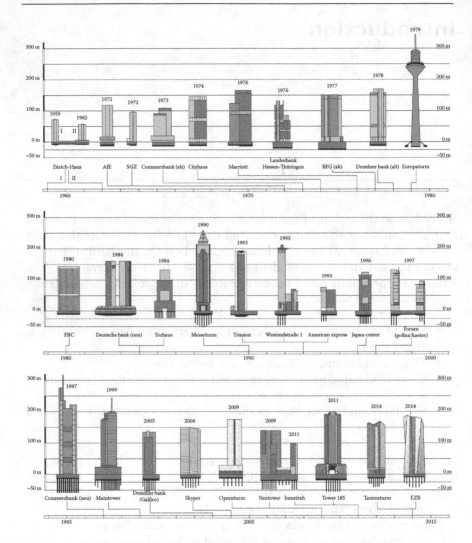

Figure 1.1 Development of high-rise buildings in Frankfurt am Main, Germany.

Chapter 2

Basics of geotechnical analysis

2.1 SOIL–STRUCTURE INTERACTION

To ensure the stability and the serviceability of any project with sufficient certainty, the interaction between the subsoil and the bearing structure generally has to be considered. Therefore, an accurate theoretical model for the description of this interaction is necessary. At the junction between structural engineering and geotechnical engineering, the soil-structure interaction is of enormous importance [1].

For a realistic and correct theoretical model of the three-dimensional and often time-variant soil–structure interaction, the following have to be taken into account:

* Modeling of the structure and its mechanical behavior
* Modeling of the soil and its mechanical behavior in relation to the multiphase material soil
* Determination of the contact behavior between subsoil and structure

During the design process for the various elements of a structure, different theoretical models can be used for considering the soil–structure interaction. The soil is not only a stabilizing or a loading element, but also, in combination with other construction elements, it is a hybrid bearing system. On the one hand, the loads from the structures create the main limit state for the analysis of the stability of the structures and the foundation elements. On the other hand, the settlements and differential settlements in the soil cause the main limit state for the analysis of the serviceability of the structures.

The soil is a part of the static system, but it may also add loading on structures due to its own weight. Therefore, two construction types have to be distinguished:

* Foundation systems (spread foundations, deep foundations, etc.), which are sustained by the soil

3

- Support structures (retaining walls, tunnels, etc.), which resist the soil

For analysis of stability and serviceability, the material behavior of the soil and the structure has to be taken into account. Often, elastoplastic, nonlinear constitutive laws are used, depending on the stress level and the velocity of the application of the load.

Moreover, the time-dependent effect of soil–structure interaction must also be considered [2–5]. This effect is caused by

- Successive construction works involving structural changes and load changes
- Changes of the rigidity
- Shifts of the load center
- Successive excavation or deconstruction works
- Successive installations, modifications or removals of anchors, stiffeners, and so on.
- Changes of the material behavior (creep, shrinkage, consolidation)

Thus, continuously changing static systems occur during the construction phases. Figure 2.1 shows the development of the qualitative deformations and loads in the bottom of an excavation during a construction project and the resulting deformations of construction elements that are built at various times [1,6].

2.2 ANALYSIS ACCORDING TO EUROCODE 7 (EC 7)

After 30 years of development and implementation of the Eurocodes, an integrative regulation of the analysis in civil engineering disciplines was achieved. The Eurocodes are based on the principle of the partial safety factor concept, which replaces the global safety factor concept.

Eurocode 7 (EC 7), consisting of two parts, was developed for geotechnical engineering. The first part [7] contains general regulations, the second part [8] contains the field investigations.

In Germany, for example, EC 7 was established with the following regulations in December 2012:

- DIN EN 1997-1 [9]
- DIN EN 1997-1/NA [10]
- DIN EN 1997-2 [11]
- DIN EN 1997-2/NA [12]
- DIN 1054 [13]
- DIN 1054/A1 [14]
- DIN 4020 [15]

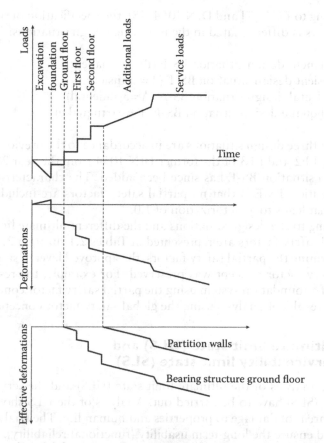

Figure 2.1 Loads and deformations during the construction process. (From Katzenbach et al., *Handbuch für Bauingenieure: Technik, Organisation und Wirtschaftlichkeit.* Springer-Verlag, Heidelberg, Germany, 1471–1490, 2012; Katzenbach et al., Die Berücksichtigung und Modellierung der Interaktion zwischen Baugrund und Tragwerk ist für die Standsicherheit und Gebrauchstauglichkeit der Konstruktion von entscheidender Bedeutung. Prüfingenieur, Vogel Druck und Medienservice, Germany, 44–62, 2013.)

These standards were integrated in user-friendly compendiums [16], which are separated in the different parts 1 [17] and 2 [18].

2.2.1 Design situations

Eurocode 0 distinguishes the following design situations [19–23]:

- Permanent
- Transient
- Accidental
- Earthquake

According to EC 7 [7] and DIN 1054 [13], the specification of the partial safety factors is differentiated in the following design situations:

- Permanent design situation BS-P (P=permanent)
- Transient design situation BS-T (T=transient)
- Accidental design situation BS-A (A=accidental)
- Earthquake design situation BS-E (E=earthquake)

The first three design situations are in accordance to the previous loading cases LF1, LF2 and LF3 of the former DIN 1054 from the year 2005 [24]. The design situation BS-E has since been added. The characteristic of the design situation BS-E is that no partial safety factors are included in the analysis that leads to a factorization of 1.0.

According to the design situations and the different ultimate limit states, the partial safety factors are represented in Tables 2.1 through 2.3 [13].

To determine the partial safety factors, the approved level of safety of the global safety factor concept was preserved. For example, the result of an analysis of a foundation system using the partial safety factor concept leads to similar results of analysis using the global safety factor concept.

2.2.2 Ultimate limit state (ULS) and serviceability limit state (SLS)

Generally, analysis of the ultimate limit state (ULS) and the serviceability limit state (SLS) have to be carried out. Analysis of the ULS should eliminate the threat of damage to properties and human life. The analysis of the SLS should ensure the long-term usability (functional reliability).

In geotechnical engineering as well as in other parts of civil engineering, five different limit states are defined [7,9,10,13,14,19,20–22]. Table 2.4 compares the limit states of the previously applicable DIN 1054 and the current regulations [7,9,10,13,14].

Regarding the ULS, the design values of the loading E_d are opposed to the design values of the resistance R_d of a structure or a structural element. $E_d \leq R_d$ has to be observed.

SLS problems are those that restrict the usability or function of a structure, the well-being of its inhabitants, or the appearance of a structure.

Regarding the SLS, the design value of loading E_d has to be smaller than the design value of the decisive serviceability criterion C_d. $E_d \leq C_d$ has to be observed. Usually, the partial safety factors are 1.0 for analysis of SLS.

2.2.3 Rules for combination factors

In line with the implementation of the Eurocode, the application of combination factors in geotechnical engineering was adopted. In this regard, consideration is given to the probability of a simultaneous effect of the variable

Table 2.1 Safety factors for influences and loads according to DIN 1054

Influence resp. load		Symbol	Design situation		
			BS-P	BS-T	BS-A
ULS HYD and UPL	Destabilizing permanent influences[a]	$\gamma_{G,dst}$	1.05	1.05	1.00
	Stabilizing permanent influences	$\gamma_{G,stb}$	0.95	0.95	0.95
	Destabilizing changeable influences	$\gamma_{Q,dst}$	1.50	1.30	1.00
	Stabilizing changeable influences	$\gamma_{Q,stb}$	0	0	0
	Flow stress in favorable soil	γ_H	1.35	1.30	1.20
	Flow stress in unfavorable soil	γ_H	1.80	1.60	1.35
EQU	Unfavorable permanent loads	$\gamma_{G,dst}$	1.10	1.05	1.00
	Favorable permanent loads	$\gamma_{G,stb}$	0.90	0.90	0.95
	Unfavorable changeable loads	γ_Q	1.50	1.25	1.00
STR and GEO-2	Loads resulting from permanent influences in general[a]	γ_G	1.35	1.20	1.10
	Loads resulting from favorable permanent influences[b]	$\gamma_{G,inf}$	1.00	1.00	1.00
	Loads resulting from permanent influences of the at-rest earth pressure	$\gamma_{G,E0}$	1.20	1.10	1.00
	Loads resulting from unfavorable changeable influences	γ_Q	1.50	1.30	1.10
	Loads resulting from favorable changeable influences	γ_Q	0	0	0
GEO-3	Permanent loads[a]	γ_G	1.00	1.00	1.00
	Unfavorable permanent loads	γ_Q	1.30	1.20	1.00
SLS	Permanent influences resp. loads	γ_G		1.00	
	Changeable influences resp. loads	γ_Q		1.00	

Source: Deutsches Institut für Normunge.V. (2010): *DIN 1054 Subsoil—Verification of the Safety of Earthworks and Foundations—Supplementary Rules to DIN EN 1997-1*. Beuth Verlag, Germany, table A 2.1.

[a] Including a permanent and changeable water pressure.
[b] Only if the determination of the value of the tensile load considers a simultaneously compressive load.

loads in full size. According to the combination rules, only the main influence $Q_{k,l}$ is considered in full size combined with a simultaneous influence of different changeable loads. All further loads, the so-called accompanying loads $Q_{k,i}$ are attenuated by a combination coefficient.

2.2.4 General procedure of analysis

Despite the switch from the global safety factor concept to the partial safety factor concept, the essential procedures of analyses in the ultimate limit

Table 2.2 Safety factors for geotechnical values according to DIN 1054

				Design situation	
Soil parameters		Symbol	BS-P	BS-T	BS-A
HYD and UPL	Friction coefficient tan(φ') of the drained soil and friction coefficient tan(φ_u) of the undrained soil	$\gamma_{\varphi'}, \gamma_{\varphi,u}$	1.00	1.00	1.00
	Cohesion c' of the drained soil and shear strength c_u of the undrained soil	$\gamma_{c'}, \gamma_{cu}$	1.00	1.00	1.00
GEO-2	Friction coefficient tan(φ') of the drained soil and friction coefficient tan(φ_u) of the undrained soil	$\gamma_{\varphi'}, \gamma_{\varphi,u}$	1.00	1.00	1.00
	Cohesion c' of the drained soil and shear strength c_u of the undrained soil	$\gamma_{c'}, \gamma_{cu}$	1.00	1.00	1.00
GEO-3	Friction coefficient tan(φ') of the drained soil and friction coefficient tan(φ_u) of the undrained soil	$\gamma_{\varphi'}, \gamma_{\varphi,u}$	1.25	1.15	1.10
	Cohesion c' of the drained soil and shear strength c_u of the undrained soil	$\gamma_{c'}, \gamma_{cu}$	1.25	1.15	1.10

Source: Deutsches Institut für Normunge.V. (2010): *DIN 1054 Subsoil—Verification of the Safety of Earthworks and Foundations—Supplementary Rules to DIN EN 1997-1.* Beuth Verlag, Germany, table A 2.2.

Table 2.3 Safety factors for resistance according to DIN 1054

			Design situation	
Resistance		Symbol	BS-P	BS-T	BS-A
STR and GEO-2	**Soil resistances**				
	Earth resistance and base failure resistance	$\gamma_{R,e}, \gamma_{R,v}$	1.40	1.30	1.20
	Sliding resistance	$\gamma_{R,h}$	1.10	1.10	1.10
	Pile resistances determined by static and dynamic pile load tests				
	Base resistance	γ_b	1.10	1.10	1.10
	Shaft resistance (pressure)	γ_s	1.10	1.10	1.10
	Total resistance (pressure)	γ_t	1.10	1.10	1.10
	Shaft resistance (tension)	$\gamma_{s,t}$	1.15	1.15	1.15
	Pile resistances on the basis of experience				
	Pressure piles	$\gamma_b, \gamma_s, \gamma_t$	1.40	1.40	1.40
	Tension piles (only in exceptional cases)	$\gamma_{s,t}$	1.50	1.50	1.50
	Pull-out resistances				
	Soil resp. rock nails	γ_a	1.40	1.30	1.20
	Grout body of grouted anchors	γ_a	1.10	1.10	1.10
	Flexible reinforcement elements	γ_a	1.40	1.30	1.20

Source: Deutsches Institut für Normunge.V. (2010): *DIN 1054 Subsoil—Verification of the Safety of Earthworks and Foundations—Supplementary Rules to DIN EN 1997-1.* Beuth Verlag, Germany, table A 2.3.

Table 2.4 Comparison of the acronym of the ultimate limit state design according to DIN 1054 [24] and EC 7 [7,9,10,13,14]

DIN 1054:2005-01		EC 7-1 and DIN 1054:2010-12	
Description	Acronym	Description	Acronym
Loss of static equilibrium	GZ 1A	Loss of static equilibrium/tilting	EQU (equilibrium)
		Buoyancy (analysis as GZ 1A)	UPL (uplift)
		Hydraulic failure (analysis as GZ 1A)	HYD (hydraulic)
Failure of constructions and constructions elements due to failure in the construction or in the supporting soil	GZ 1B	Failure of the structure or its elements	STR (structural)
		Failure of soil (analysis as GZ 1B)	GEO-2
Limit state of the loss of overall stability	GZ 1C	Failure of soil (analysis as GZ 1C (Fellenius-rule))	GEO-3

and serviceability limit states were preserved. All analyses, except GEO-3, are based to a factorization of the forces by partial safety factors. The characteristic values of the loads (e.g., loads from superstructure, earth pressure, water pressure) lead to characteristic values of the forces and stresses (e.g., stresses under foundation raft at analysis of the base failure or sliding) and lead to characteristic values of the resistances (resistance of base failure or resistance against sliding).

In the limit state of GEO-3, analysis 3 is implemented with attenuated shear parameters. Consequently, the calculation of the forces and stresses is based on the design values.

All characteristic values of forces and stresses of any kind of structure are essential as input parameters for the geotechnical analyses. Figure 2.2 shows the general procedure of analysis.

2.2.5 Geotechnical categories

The determination of minimum requirements in relation to the geotechnical investigations, analyses, and monitoring depends on the geotechnical categories GC 1 to GC 3 [7,8]. The classification to one of the three geotechnical categories has to be carried out before the planning of the soil investigation program. The criterion that causes the highest geotechnical category is decisive. If necessary, the classification has to be adopted during the planning and construction phases.

The geotechnical category GC 1 comprises construction projects with a low level of difficulty with regard to the soil and the structures:

- Simple, predictable soil conditions (horizontal or slightly inclined surfaces, and based on local experience, stable soils with little settlement)

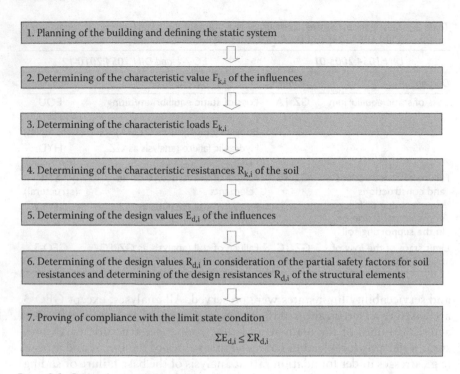

1. Planning of the building and defining the static system

2. Determining of the characteristic value $F_{k,i}$ of the influences

3. Determining of the characteristic loads $E_{k,i}$

4. Determining of the characteristic resistances $R_{k,i}$ of the soil

5. Determining of the design values $E_{d,i}$ of the influences

6. Determining of the design values $R_{d,i}$ in consideration of the partial safety factors for soil resistances and determining of the design resistances $R_{d,i}$ of the structural elements

7. Proving of compliance with the limit state conditon
$$\Sigma E_{d,i} \leq \Sigma R_{d,i}$$

Figure 2.2 **General procedure of the analysis.**

- Groundwater level is below the excavation or the foundation level
- Non–settlement-sensitive structures with spread foundations and vertical column loads up to 250 kN and strip loads up to 100 kN/m, such as family houses, single-floor halls, or garages
- Structures where analysis of stability regarding earthquake loads are not needed, in accordance to [25]
- Neighboring buildings, infrastructures, pipes, and so on are not endangered by the stability or usability of the new structure or the necessary construction processes

Examples of structures that are classified to the geotechnical category GC 1:

- Single and strip foundations where the requirements for the simplified procedures of analysis are fulfilled
- Foundation rafts under well-braced structures with a maximum of two levels above surface
- Retaining structures up to 2 m in height and without high loads rearward against the wall

The geotechnical category GC 2 comprises construction projects with conditions of a medium level of difficulty with regard to the interaction between the soil and the structure:

- Average soil conditions that are not included in GC 1 or GC 3
- Free groundwater level that is higher than the excavation or foundation level
- Groundwater flow or dewatering that could be implemented by common measures without injurious impact the neighborhood
- Common buildings and civil engineering structures on single foundations, strip foundations, foundation rafts, or pile foundation systems
- Structures, where analysis of stability regarding earthquake loads are needed, in accordance with [25]
- Construction projects that do not have an injurious impact on the neighborhood and surrounding area due to robust constructions (e.g., non-permeable retaining systems with stiff bracings)

Examples of structures that are classified to the geotechnical category GC 2:

- Common single foundations, strip foundations, and foundation rafts, which are not included in GC 1 or GC 3
- Pipe ditches and trenches up to a depth of 5 m
- Retaining structures up to a height of 10 m
- Construction projects that require an analysis of safety against buoyancy of non-anchored construction
- Construction projects that require an analysis of safety against hydraulic failure

The geotechnical category GC 3 comprises construction projects with conditions of a high level of difficulty with regard to the interaction between the soil and the structure:

- Young geological deposits with irregular stratifications respective to unsettled geological formations
- Soils that tend to creep, flow, heave, or shrink
- Cohesive soils, where the residual shear strength could be decisive
- Cohesive soils without sufficient ductility, for example, structure-sensitive sea clay
- Soft, organic, and organogenic soils with a large thickness
- Rock materials that tend to decay or dissolve, or variable solid rocks
- Rock that is unfavorable, crossed by interference zones and partition surfaces
- Mining sinkhole areas, or areas with collapsed sinkholes or unsecured hollows in the underground
- Uncontrolled backfilling

- Confined groundwater
- Structures with high security requirements or high sensitivity for deformations
- Structures with exceptional load combinations that are decisive for the foundation
- Structures that are loaded by a water pressure height of more than 5 m
- Facilities and construction projects that change the groundwater level temporarily or permanently in combination with risks for neighboring buildings
- Structures that belong to the categories of significance III and IV in accordance to [25], which requires an analysis of stability with regard to earthquake loads
- Structures or construction projects where the observational method has to be applied in addition to the common analyses of stability and serviceability

Examples of structures that are classified to the geotechnical category GC 3:

- Construction projects with a distinctive soil–structure interaction, for example, mixed foundations and foundation rafts
- Structures with significant and variable water pressure influences
- Structures with extremely high loads, for example, single loads of 10 MN and more
- Foundations for bridges with large spans, for example, 40 m, and with static indeterminate supported superstructures that would be influenced by constraining forces owing to different settlements of the supporting pillars and abutments, as well as integral bridges
- Machine foundations with high dynamic loads
- Foundations for towers, transmitter masts, and industrial chimneys
- Extended raft foundations based on a soil with various degrees of stiffness in groundview
- Foundations in the vicinity of existing buildings, if the conditions according to [26] do not apply
- Structures with different foundation levels, or with different foundation elements
- Combined Pile-Raft Foundations (CPRF)
- Caisson foundations combined with compressed air
- Underground constructions, tunnels, studs, and shafts in soil or fractured rock
- Nuclear facilities
- Offshore constructions
- Chemical plants and constructions where dangerous substances are produced, stored, or handled
- Special methods and techniques, for example, diaphragm walls and jet grouting

- Retaining structures with a height of more than 10 m, or excavations in soft soils

A comprehensive review of examples for the classification to geotechnical categories is given in [16].

2.3 SOIL INVESTIGATION ACCORDING TO EUROCODE 7 (EC 7)

By the implementation of EC 7, the basic regulations for soil investigation all over Europe are defined. The national application of the regulations, for example, in Germany, is given by [11,12,15] and is summarized in [17]. A sufficient knowledge of the soil and groundwater conditions is essential for a secure and economical design of foundation systems. The difficulty of soil investigations is that even with an extensive investigation program, less than 0.1 per mill of the volume of the soil influenced by the structure is determined, as shown in Figure 2.3 [27,28]. Also, the interpretation and

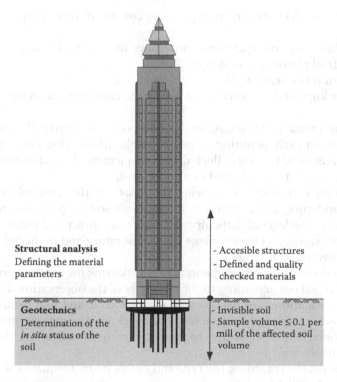

Structural analysis
Defining the material
parameters

- Accesible structures
- Defined and quality checked materials

Geotechnics
Determination of the
in situ status of the
soil

- Invisible soil
- Sample volume ≤ 0.1 per mill of the affected soil volume

Figure 2.3 Determination of the material parameters in structural analysis and in geotechnics.

evaluation of the investigation results can vary; this is shown in Figure 2.4 by two direct ground explorations, for example, core drillings. The stratigraphy between the two explorations could either be continuous soil changing or a shifting soil changing.

2.3.1 Soil investigation program

An adequate soil investigation program is adapted to address the complexities of each geotechnical category. An investigation program contains different measures: *in situ* measures taken at the project area, and measures determined by geotechnical laboratory tests, as shown in Figure 2.5 [29]. For construction projects that are classified to the geotechnical category GC 2 or GC 3, a geotechnical expert has to be involved.

The soil investigation *in situ* is divided into direct and indirect investigation measures. Direct investigation measures are, for example, testpits, core drillings and *in situ* field tests. Indirect investigation measures are, for example, driving and cone penetration tests, as well as geophysical measuring methods. The soil mechanical parameters are determined by laboratory tests.

An adequate soil investigation program consists of three parts:

- Preliminary investigations concerning the location and preliminary draft of planning and design
- Main investigations
- Checkups and measurements during the construction phase

The preliminary investigations are necessary to verify the proposed location at an early planning stage. Available information about soil and groundwater conditions is collected and complemented by additional investigations, which are conducted in a rough grid.

The main investigations provide the basis for the detailed planning, design, tendering, and construction. The main soil investigation program is adapted to the level of difficulty and comprises direct and indirect investigation measures and experiments in the laboratory and in the field with a suitable investigation grid.

The checkups and measurements take place during the construction phase and are carried out according to the principles of the observational method. These investigations are necessary if unpredictable soil and groundwater conditions are expected or detected. The checkups and measurements during the construction phase aim to verify the main investigation, the analysis, and the design.

Further details regarding the type and extent of an adequate soil investigation are described in [30].

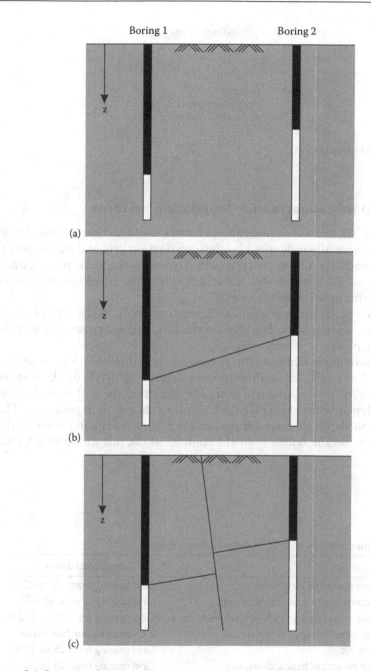

Figure 2.4 Example of the results of the soil investigation by borings (a) and its possible interpretation (b and c).

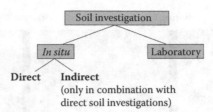

Figure 2.5 Soil investigation.

2.3.2 Soil investigation for foundation systems

The content of the soil investigations for the construction of foundation systems is essentially influenced by the investigation grid and the investigation depth and depends on the type of the structure, the foundation system, and the expected stratigraphy. Table 2.5 shows the investigation grid of different structures according to EC 7.

The depth of the soil investigation depends not only on the type of structure and the stratigraphy, but also on the foundation system and its geometric dimensions.

The investigation depth z_a (in meters) of spread foundations is shown in Figures 2.6 and 2.7. The investigation depth z_a depends on the smaller width b_F or b_B of the construction. The depth of the soil investigation of dams depends on the height h in [m] as shown in Figure 2.8. The investigation depth of deep foundations is reliant on the diameter of the pile toe resp. on the width b_g of the contour of the pile group or a CPRF (Figure 2.9).

Table 2.5 Investigation grid depending on the structure

Structure	Horizontal distance
Buildings and industrial structures	Grid spacing from 15 to 40 m
Extensive constructions (warehouses, etc.)	Grid spacing of max. 60 m
Line constructions (streets, railways, channels, etc.)	Grid spacing from 20 to 200 m
Special structures (bridges, chimneys, etc.)	2 to 6 points per foundation
Dams, weirs, etc.	Grid spacing from 25 to 75 m
Large water retention basins, dams, etc.	Grid spacing from 25 to 50 m

Source: Katzenbach, R.; Schuppener, B.; Weidle, A.; Ruppert, T. (2011): Grenzzustandsnachweise in der Geotechnik nach EC 7-1. *Bauingenieur* 86, Heft 7/8, Springer VDI Verlag, Düsseldorf, Germany, 356–363.

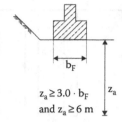

$z_a \geq 3.0 \cdot b_F$
and $z_a \geq 6$ m

Figure 2.6 Investigation depth for strip and single foundations.

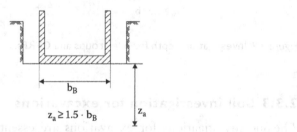

$z_a \geq 1.5 \cdot b_B$

Figure 2.7 Investigation depth for raft foundations.

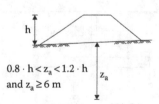

$0.8 \cdot h < z_a < 1.2 \cdot h$
and $z_a \geq 6$ m

Figure 2.8 Investigation depth for dams.

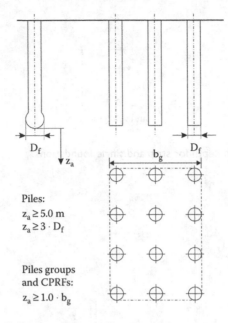

Figure 2.9 Investigation depth for pile groups and CPRFs.

2.3.3 Soil investigation for excavations

The soil investigations for excavations are essentially influenced by the excavation depth in [m] and the embedment depth of the retaining system in [m] and depend on the type of structure and the expected stratigraphy. Two distinct groundwater situations are constituted in Figures 2.10 and 2.11, respectively:

- Groundwater level beneath the excavation
- Groundwater level above the excavation

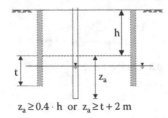

$z_a \geq 0.4 \cdot h$ or $z_a \geq t + 2\,m$

Figure 2.10 Investigation depth for excavations when groundwater level is below the excavation.

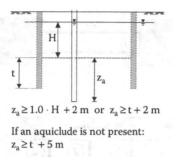

$z_a \geq 1.0 \cdot H + 2\,m$ or $z_a \geq t + 2\,m$

If an aquiclude is not present:
$z_a \geq t + 5\,m$

Figure 2.11 Investigation depth for excavations when groundwater level is above the excavation.

2.4 GUARANTEE OF SAFETY AND OPTIMIZATION BY THE FOUR-EYE PRINCIPLE

The large number of accidents in construction projects in recent years shows that, for safety aspects, an independent supervision and monitoring system is necessary during planning, design, and construction. To guarantee public safety, the four-eye principle is a vital for the verification of analyses and designs by a publicly certified independent expert [31,32]. The Association for Urban Development, Building and Housing of the Federal Republic of Germany established the four-eye principle in the national building regulations [33]. The detailed description of the qualifications and responsibilities of a publicly certified independent expert is given in [34]. Publicly certified independent experts in the following fields are required:

- Structural engineering
- Fire prevention
- Technical facilities and installations
- Geotechnical engineering

The publicly certified independent experts verify and certify compliance to the current standards and regulations in their specific field of work. Publicly certified independent experts for geotechnics verify and certify the completeness and accuracy of the soil investigation (stratigraphy, soil parameters, groundwater conditions, bearing capacity, stiffness etc.) and the planning, design and construction of foundation systems, retaining structures, tunnels, and so on.

The importance of the four-eye principle regarding geotechnics becomes apparent in [35]: All structural and environmental constructions are concerned with soil and rock. An accurate description, evaluation, and handling of the soil and rock parameters in planning, design, and construction is essential and is, in difficult cases, very important during the service

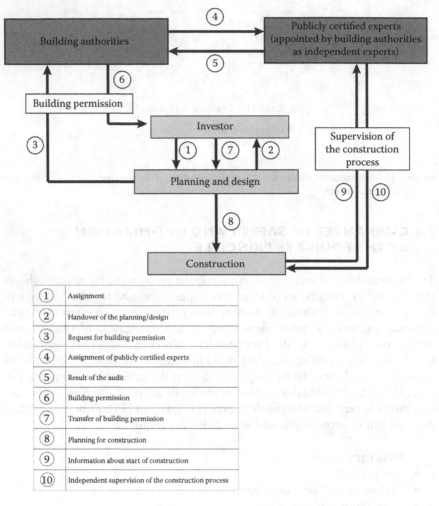

Figure 2.12 Four-eye principle.

phase after construction. In [36] the importance of the four-eye principle is demonstrated by several case studies of the geotechnical engineering practice.

The four-eye principle of civil engineering consists of three major parts, which are shown in Figure 2.12. Investors, experts for planning and design, and construction companies belong to the first part. Planning and design are based on the current standards and regulations and are parts of the request for the building permission. The building authorities are the second part, and they independently check compliance of the planning to the building law. The building authorities are responsible for the independent supervision of all legal aspects. The third part consists of the publicly certified independent experts. They are responsible for the independent supervision of all engineering aspects during planning, design, and construction.

2.5 OBSERVATIONAL METHOD

The observational method is a verification procedure that was introduced by the building authorities. Compared to other construction materials like concrete or steel, this method takes into account the difficulty and probably insufficient specifications and descriptions of soil material behavior. Furthermore, possible irregularities between the soil and rock mechanic parameters in the theoretical models and the soil and groundwater conditions *in situ* may differ [37,38]. This method is of importance for the engineering as well as for the legal aspects of a project [39].

The observational method is a combination of geotechnical investigations and analyses with a metrological supervision during the construction phase and, if necessary, during the service phase. Critical situations have to be controlled by suitable technical measures. The observational method is a precise monitoring procedure to verify the soil and rock mechanical modeling, as well as the quality and safety during the construction phase (Figure 2.13).

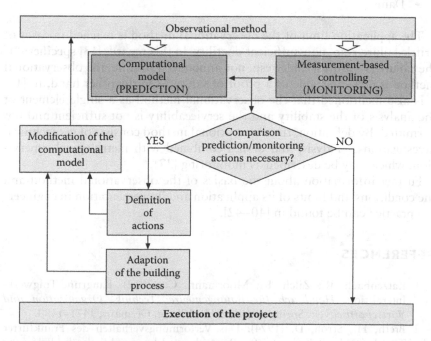

Figure 2.13 Observational method. (From Deutsches Institut für Normung e.V, *DIN 4123 Excavations, Foundations and Underpinnings in the Area of Existing Buildings.* Beuth Verlag, Berlin, Germany, 2013.)

The application of the observational method leads to a verification of the usability and the validation of the theoretical models and to quality assurance during the construction phase. Unexpected measurement data sometimes lead to dispute between the different participants of the construction project with respect to which theoretical model to apply. In these discussions, safety issues must be considered.

In accordance to the current technical standards and regulations [9–15], the observational method is state of the art for construction projects with significant geotechnical difficulties (geotechnical category GC 3). Examples for these construction projects are

- Construction projects with distinct soil–structure interaction, for example, high-rise buildings, mixed foundations, foundation rafts, CPRFs, deep excavations
- Complex interactions between soil, retaining structures, and adjacent buildings
- Structures with significant and variable influences of water pressure, for example, trough structures or water wings in tidal areas
- Construction projects where the stability could be reduced due to pore water pressure
- Tunnels
- Dams

The application limit of the observational method is reached in cases of brittle failure resp. non-sufficient ductility. In this regard, [13] specifies: "If the failure is unforeseeable resp. not announced in time, the observational method is not applicable as a proof of safety." On the other hand, in [13] it is also mentioned that the observational method as a single element of the analysis of the stability and the serviceability is not sufficient and not permitted. By definition, the observational method consists of geotechnical investigations, analysis, and design combined with metrological supervision, which may be described as monitoring [37].

Further information about the basics of the observational method and the conditions and limits of its application and implementation in engineering practice can be found in [40–42].

REFERENCES

1. Katzenbach, R.; Zilch, K.; Moormann, C. (2012): Baugrund-Tragwerk-Interaktion. *Handbuch für Bauingenieure: Technik, Organisation und Wirtschaftlichkeit*. Springer-Verlag, Heidelberg, Germany, 1471–1490.
2. Breth, H.; Stroh, D. (1974): Das Verformungsverhalten des Frankfurter Tons beim Aushub einer tiefen Baugrube und bei anschließender Belastung durch ein Hochhaus. 13. *Baugrundtagung der Deutschen Gesellschaft für Geotechnik in Frankfurt am Main*, Germany, 51–70.

3. Katzenbach, R.; Leppla, S.; Seip, M. (2011): Das Verformungsverhalten des Frankfurter Tons infolge Baugrundentlastung. *Bauingenieur* 86, May, Springer VDI-Verlag, Düsseldorf, Germany, 233–240.
4. Katzenbach, R., Leppla, S.; Krajewski, W. (2014): Numerical analysis and verification of the soil–structure interaction in the course of large construction projects in inner cities. *International Conference on Soil–Structure Interaction: Underground Structures and Retaining Walls*, 16–18 June, St. Petersburg, Russia, 28–34.
5. Katzenbach, R.; Leppla, S. (2014): Deep foundation systems for high-rise buildings in difficult soil conditions. *Geotechnical Engineering Journal of the SEAGS & AGSSEA*, Vol. 45, No. 2, 115–123.
6. Katzenbach, R.; Bergmann, C.; Leppla, S.; Kurze, S.; Seip, M. (2013): Die Berücksichtigung und Modellierung der Interaktion zwischen Baugrund und Tragwerk ist für die Standsicherheit und Gebrauchstauglichkeit der Konstruktion von entscheidender Bedeutung. Prüfingenieur, Vogel Druck und Medienservice, Höchberg, Germany, 44–62.
7. CEN European Committee of Standardisation (2008): Eurocode 7: Geotechnical design—Part 1: General Rules.
8. CEN European Committee of Standardisation (2008): Eurocode 7: Geotechnical design—Part 2: Ground Investigation and Testing.
9. Deutsches Institut für Normung e.V. (2014): DIN EN 1997-1 Eurocode 7: Geotechnical design—Part 1: General Rules. Beuth Verlag, Berlin, Germany.
10. Deutsches Institut für Normung e.V. (2010): DIN EN 1997-1/NA National Annex—Nationally determined parameters—Eurocode 7: Geotechnical Design—Part 1: General Rules. Beuth Verlag, Berlin, Germany.
11. Deutsches Institut für Normung e.V. (2010): DIN EN 1997-2 Eurocode 7: Geotechnical Design—Part 2: Ground Investigation and Testing. Beuth Verlag, Berlin, Germany.
12. Deutsches Institut für Normung e.V. (2010): DIN EN 1997-2/NA National Annex—Nationally determined parameters—Eurocode 7: Geotechnical Design—Part 2: Ground Investigation and Testing. Beuth Verlag, Berlin, Germany.
13. Deutsches Institut fürNormunge. V. (2010): DIN 1054 Subsoil—Verification of the Safety of Earthworks and Foundations—Supplementary Rules to DIN EN 1997-1. Beuth Verlag, Berlin, Germany.
14. Deutsches Institut für Normunge. V. (2012): DIN 1054 Subsoil—Verification of the Safety of Earthworks and Foundations—Supplementary Rules to DIN EN 1997-1:2010; Amendment A1. Beuth Verlag, Berlin, Germany.
15. Deutsches Institut für Normunge. V. (2010): DIN 4020 Geotechnical Investigations for Civil Engineering Purposes: Supplementary Rules to DIN EN 1997-2. Beuth Verlag, Berlin, Germany.
16. Katzenbach, R.; Schuppener, B.; Weidle, A.; Ruppert, T. (2011): Grenzzustandsnachweise in der Geotechnik nach EC 7-1. *Bauingenieur* 86, Heft 7/8, Springer VDIVerlag, Düsseldorf, Germany, 356–363.
17. Deutsches Institut für Normung e.V. (2011): Handbuch Eurocode 7, Geotechnische Bemessung, Band 1: Allgemeine Regeln. Beuth Verlag, Berlin, Germany.
18. Deutsches Institut für Normung e.V. (2011): Handbuch Eurocode 7, Geotechnische Bemessung, Band 2: Erkundung und Untersuchung. Beuth Verlag, Berlin, Germany.

19. CEN European Committee of Standardisation (2002): Eurocode 0: Basis of structural design.
20. Deutsches Institut für Normung e.V. (2010): DIN EN 1990 Eurocode: Basis of Structural Design. Beuth Verlag, Berlin, Germany.
21. Deutsches Institut für Normung e.V. (2010): DIN EN 1990/NA National Annex—Nationally Determined Parameters—Eurocode: Basis of Structural Design. Beuth Verlag, Berlin, Germany.
22. Deutsches Institut für Normung e.V. (2012): DIN EN 1990/NA/A1 National Annex—Nationally Determined Parameters—Eurocode: Basis of Structural Design Amendment A1. Beuth Verlag, Berlin, Germany.
23. Deutsches Institut für Normung e.V. (2011): Handbuch Eurocode 0, Grundlagen der Tragwerksplanung. Beuth Verlag, Berlin, Germany.
24. Deutsches Institut für Normunge. e.V. (2005): DIN 1054 Subsoil—Verification of the Safety of Earthworks and Foundations. Beuth Verlag, Berlin, Germany.
25. Deutsches Institut für Normunge.V. (2011): DIN EN 1998–5/NA National Annex-Nationally Determined Parameters—Eurocode 8: Design of Structures for Earthquake Resistance—Part5: Foundations, Retaining Structures and Geotechnical Aspects. Beuth Verlag, Berlin, Germany.
26. Deutsches Institut für Normung e.V. (2013): DIN 4123 Excavations, Foundations and Underpinnings in the Area of Existing Buildings. Beuth Verlag, Berlin, Germany.
27. Katzenbach, R.; Schmitt, A.; Turek, J. (1999): Cooperation between the geotechnical and structural engineers: Experiences from projects in Frankfurt. COST Action 7, Soil–Structure Interaction in Urban Civil Engineering, 1–2 October, Thessaloniki, Greece, 53–65.
28. Katzenbach, R.; Weidle, A.; Kurze, S. (2012): Baugrund und Grundwasser Erkundungsproblematik, Baugrundrisiko und technische Risiken. 39. Baurechtstagung der Arge Baurecht des Deutschen Anwaltsvereins, 16.-17. March, Berlin, Germany.
29. Eitner, V.; Katzenbach, R.; Stölben, F. (2002): Geotechnical investigation and testing: An outlook on European and international standardization. Foundation Design Codes and Soil Investigation on view of International Harmonization and Performance, Honjo, Kusakabe, Matsui, Kouda & Pokharel (Hrsg.), Swets&Zeitlinger, Lisse, the Netherlands, 211–215.
30. Pulsfort, M. (2012): *Grundbau, Baugruben und Gründungen. Handbuch für Bauingenieure: Technik, Organisation und Wirtschaftlichkeit*, Springer-Verlag, Heidelberg, Germany, 1568–1639.
31. Katzenbach, R.; Kinzel, J. (2001): Das Vier-Augen-Prinzip bei Baugrundgutachten. Der Prüfingenieur, Nr. 18, Vogel Verlag, Würzburg, Germany, 28–38.
32. Katzenbach, R.; Boley, C.; Moormann, C.; Rückert, A. (1999): Rechtsrelevante Sicherheitsaspekte in der Geotechnik. 1. Darmstädter Baurechts-Kolloquium, 14. January, Mitteilungen des Institutes und der Versuchsanstalt für Geotechnik der Technischen Universität Darmstadt, Germany, Heft 43, 71–96.
33. Bauministerkonferenz (2012): Musterbauordnung (MBO). Germany.
34. Bauministerkonferenz (2012): Muster-Verordnung über die Prüfingenieure und Prüfsachverständigen nach § 85 Abs. 2 MBO (H-PPVO). Germany.

35. Floss, R.; Gudehus, G.; Katzenbach, R.: Zur Position der Geotechnik als zentrale Disziplin des Bauingenieurwesens. Geotechnik 23, Nr. 1, VGE Verlag, Essen, Germany, 12–15.
36. Katzenbach, R.; Leppla, S.; Weidle, A.; Werner, A. (2011): Das Vier-Augen-Prinzip in der Geotechnik: Der Prüfsachverständige für Erd- und Grundbau. Geotechnik-Kolloquium anlässlich 60. Geburtstag von Prof. Dr.-Ing. Dietmar Placzek, 26. May, Universität Duisburg-Essen, Germany, 255–267.
37. Katzenbach, R.; Gutwald, J. (2003): Interaktion in der Geotechnik: Baugrunderkundung, Bemessung, Bauausführung und Beobachtungsmethode. DIN-Gemeinschaftstagung Bemessung und Erkundung in der Geotechnik: Neue Entwicklungen im Zuge der Neuauflage der DIN 1054 und DIN 4020 sowie der europäischen Normung, 2 April, Heidelberg, Germany, 8.1–8.24.
38. Katzenbach, R.; Bachmann, G.; Ramm, H.; Waberseck, T.; Dunaevskiy, R. (2008): Monitoring of geotechnical constructions: An indispensable tool for economic efficiency and safety of urban areas. *International Geotechnical Conference*, 16 June, St. Petersburg, Russia, 695–699.
39. Katzenbach, R.; Bachmann, G. (2006): Sicherheit und Systemoptimierung durch Monitoring in der Geotechnik. 29. Darmstädter Massivbauseminar, 14–15 September, Germany, 251–265.
40. Katzenbach, R.; Bachmann, G.; Leppla, S.; Ramm, H. (2010): Chances and limitations of the observational method in geotechnical monitoring. *Danube-European Conference on Geotechnical Engineering*, 2–4 June, Bratislava, Slovakia, 13.
41. Rodatz, W.; Gattermann, J.; Bergs, T. (1999): Results of five monitoring networks to measure loads and deformations at different quay wall constructions in the port of Hamburg. *5th International Symposium on Field Measurements in Geomechanics*, 1–3 December, Singapore, 4.
42 Moormann, C. (2002): Trag- und Verformungsverhalten tiefer Baugruben in bindigen Böden unter besonderer Berücksichtigung der Baugrund-Tragwerk-Interaktion und der Baugrund-Grundwasser-Interaktion. Mitteilungen des Institutes und der Versuchsanstalt für Geotechnik der Technischen Universität Darmstadt, Germany, Heft 59.

Chapter 3

Spread foundations

Spread foundations refer to foundation components that transfer their loads to the subsoil only by normal stresses and shear stresses. Spread foundations are single foundations, strip foundations, or raft foundations. The requirement for spread foundations is the bearing capacity of the subsoil below the bottom of the foundation. If the subsoil has insufficient bearing capacity, improvement to the subsoil or alternative foundation systems are required.

Basically, the depth of the foundation level is specified to facilitate a frost-free foundation. In Germany, this is at least 80 cm below the surface. Information on the different regional frost penetration depths is contained in [1–3].

The following incidents have to be avoided during the preparation of the foundation level:

- Leaching
- Reduction of the bulk density by drifty water
- Maceration
- Cyclic freezing and unfreezing

Before the installation of the blinding concrete, the foundation level has to be checked by a geotechnical expert.

3.1 SINGLE AND STRIP FOUNDATIONS

For the excavation of single loads like columns, single foundations are used. Strip foundations are used for line loads. Both types of spread foundation can be designed with or without a reinforcement, whereby reinforced foundations should be preferred due to their greater robustness. Figure 3.1 shows the two types of foundations.

Generally, the design of single and strip foundations based on the contact pressure is sufficient. In most cases, the contact pressure can be determined by the stress trapeze method. Deformations of the subsoil and the building, as well as the soil–structure interaction, are not taken into account.

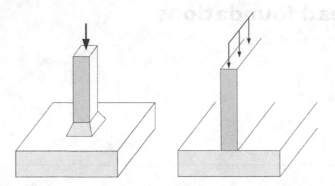

Figure 3.1 Single and strip foundation.

3.2 RAFT FOUNDATIONS

Raft foundations are used when the load grid is dense and the deformations of the subsoil and the construction have to be homogenized. Raft foundations can be used as a part of a so-called white trough, or in combination with an additional sealing system (e.g., bitumen layers) to prevent groundwater influx [4–7].

The thickness of the reinforcedconcrete slab depends on the bending moment, as well as on the punching (concentrated loads). Increasing the slab thickness or arranging concrete haunches can avoid shear reinforcements. To prevent groundwater influx or to repel weather conditions, the crack width of the concrete has to be limited. In any case, the installation of construction joints, expansion joints and settlement joints has to be planned precisely and supervised during the construction phase.

3.3 GEOTECHNICAL ANALYSIS

3.3.1 Basics

Two different theoretical models are used for the geotechnical analysis of the SLS and the ULS. For the analysis of the stability limit state (SLS), a linear elastic material behavior of the subsoil is considered. In contrast, for the design of the ultimate limit state (ULS), a rigid-plastic material behavior of the subsoil is considered. This issue of spread foundations is explained in Figure 3.2.

According to the technical standards and regulations, the following incidents have to be analyzed [8–11]:

* Overall stability
* Sliding

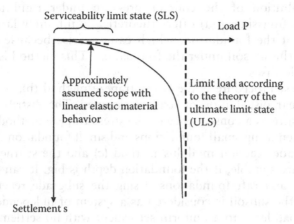

Figure 3.2 Load-settlement curve for spread foundations.

- Base failure
- Collective failure of soil and structure
- Punching, compressing
- Structural failure as a result of foundation movement
- Large settlements
- Large uplift as a result of frost
- Unacceptable vibrations

If spread foundations are located in the area of embankments, an analysis of the slope failure is necessary. Every possible failure mechanism (slip circles, complex rupture mechanisms) has to be considered [12–14].

In simple cases and under certain conditions, the geotechnical analysis of spread foundations can be done on the basis of standard table values. The standard table values take into account the analysis of safety against failure and harmful settlements [10].

3.3.2 Distribution of the contact pressure

The knowledge of the distribution of the contact pressure is the basis for the analysis of spread foundations. The following calculation procedures are available [15,16].

    a. Distribution of the contact pressure under rigid foundations according to Boussinesq [17]
    b. Stress trapeze method
    c. Subgrade reaction modulus method
    d. Stiffness modulus method
    e. Numerical methods, for example, finite element method

The distribution of the contact pressure under rigid foundations according to Boussinesq (a) offers theoretically infinitely large tensions at the edge of the foundation, which cannot arise because of transfer processes in the subsoil under the foundation. This method is applicable only in simple cases.

The simplest procedure is the stress trapeze method (b), because there is only a linear distribution of stresses assumed. The distribution of the contact pressure as a consequence of the stress trapeze method is a useful approach when using small foundations and small foundation depths.

The subgrade reaction modulus method (c) and the stiffness modulus method (d) are suitable, if the foundation depth is big. It can be used for single, strip, and raft foundations. Using the subgrade reaction modulus method, the subsoil is considered as a system of independent springs. A uniform load leads to a uniform settlement with no settlement trough. Using the stiffness modulus method, the subsoil is considered as an elastic half-space with a system of connected springs. A uniform load leads to a settlement trough. The stiffness modulus method leads to the most realistic distribution of the contact pressure.

The calculation methods (a) to (d) are approximate solutions to determine the distribution of the contact pressure below a spread foundation. These methods are usually sufficient for the analysis. The most realistic distribution of the contact pressure is given by numerical analysis because the stiffness of the structure as well the nonlinear material behavior of the subsoil can be considered.

The distribution of the contact pressure depends on the stiffness of the foundation as well as the relation between load and the stability of the subsoil [18]. The potential distributions of the contact pressure are shown in Figure 3.3. Case (a) shows the distribution of the contact pressure when the bearing capacity is exploited poorly. When the load approaches to the bearing capacity two different failure mechanisms may occur. In case (b) the load leads to a plastic hinge inside the foundation which causes a redistribution of the contact pressure. In this case the bearing capacity of the foundation depends on the rotation capacity of the plastic hinge. In case (c) the load leads to a redistribution of the contact pressure to the center of the foundation which leads to a base failure.

If the foundation has no sufficient ductility, a brittle failing follows in excess of the internal load-bearing capacity, for example, punching. A redistribution of the contact pressure will not take place.

The assumption of a constant distribution of the contact pressure leads to results on the safe side for analysis of the ULS. For analysis of the SLS, the assumption of a constant distribution of the contact pressure leads to results on the unsafe side.

Figure 3.4 shows the settlement trough, the distribution of the contact pressure, and the curve of the moment, depending on the load. With an increasing load, the constant settlements under the foundation increase

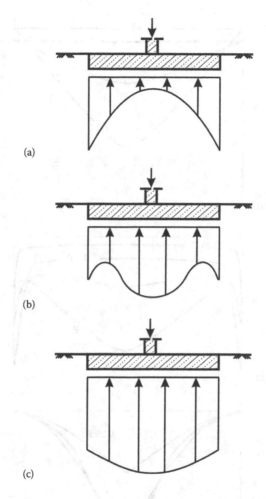

(a)

(b)

(c)

Figure 3.3 Distribution of the contact pressure under single foundations. (a) Elastic behavior of the foundation and the soil; (b) Plastic hinge in the foundation; (c) Base failure. (From Katzenbach, et al., *Baugrund-Tragwerk-Interaktion. Handbuch für Bauingenieure: Technik, Organisation und Wirtschaftlichkeit.* Springer-Verlag, Heidelberg, Germany, 1471–1490, 2012.)

strongly in the center. At the same time, the contact pressure, which is concentrated in the border area, is shifted to the center of the foundation. The bending moments are concentrated under the load.

3.3.2.1 System rigidity

For the determination of the internal force variable, the contact pressure, which depends on the proportion of the structure stiffness to the stiffness of the subsoil, needs to be analyzed.

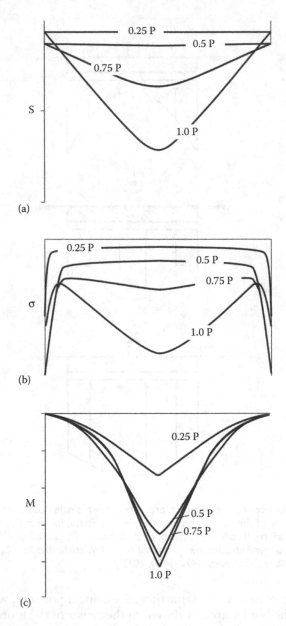

Figure 3.4 Qualitative progression of deformations and stresses of a single foundation in dependency of its loading. (a) Deformation; (b) contact pressure; (c) bending moment. (From Katzenbach, et al., *Baugrund-Tragwerk-Interaktion. Handbuch für Bauingenieure: Technik, Organisation und Wirtschaftlichkeit.* Springer-Verlag, Heidelberg, Germany, 1471–1490, 2012.)

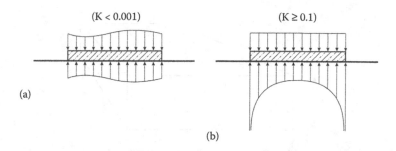

Figure 3.5 Distribution of the contact pressure for limp (a) and rigid (b) spread foundations.

Table 3.1 Differentiation between limp and rigid foundations

K ≥ 0.1	Rigid foundation
0.001 ≤ K < 0.1	Intermediate area
K < 0.001	Limp foundation

For limp spread foundations, the distribution of the contact pressure cor-respondents to the load distribution. For rigid foundations a nonlinear distri-bution of the contact pressure with high edge stresses arises (Figure 3.5). The differentiation between limp and rigid foundations is defined by the system rigidity K according to Kany, which is a value for the assessment of the inter-actions between the structure and the foundation (Equation 3.1). The dif-ferentiation is stated in Table 3.1 [16,21]. The system rigidity K is determined according to Equation 3.2. It is determined by the component height h, the length l, and the modulus of elasticity of the building material E_B, which is founded in the elastic isotropic half-space (Figure 3.6) [16–20]:

$$K = \frac{\text{structure stiffness}}{\text{subsoil stiffness}} \tag{3.1}$$

$$K = \frac{E_B \cdot I_B}{E_s \cdot b \cdot l^3} = \frac{E_B \cdot \dfrac{b \cdot h^3}{12}}{E_s \cdot b \cdot l^3} = \frac{1}{12} \cdot \frac{E_B}{E_s} \cdot \left(\frac{h}{l}\right)^3 \tag{3.2}$$

where:
E_B = modulus of elasticity of the structure [kN/m²]
I_B = geometrical moment of inertia of the spread foundation [m⁴]
E_s = oedometric modulus of the subsoil [kN/m²]
b = width of the spread foundation [m]
l = length of the spread foundation [m]
h = height of the spread foundation [m]

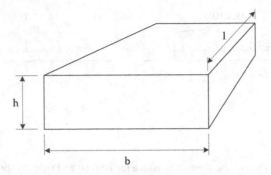

Figure 3.6 Dimensions for determining the rigidity of the system.

Circular spread foundations with a component height h and a diameter d have a system rigidity K in accordance with

$$K = \frac{1}{12} \cdot \frac{E_B}{E_S} \cdot \left(\frac{h}{d}\right)^3 \qquad (3.3)$$

For the calculation of spread foundations, normally only the rigidity of the foundation component is used to consider the rigidity of the building. The rigidity of the rising construction is considered only in special cases.

For limp spread foundations ($K < 0.001$), the settlement at the characteristic point is the same as the settlement of a rigid spread foundation (Figure 3.7). The characteristic point for rectangular foundations is at 0.74 of the half-width outward from the center. For circular spread foundations, the characteristic point is at 0.845 of the radius outward from the center.

Regardless of the load position and size, rigid spread foundations keep their forms. The distribution of the contact pressure has a strong nonlinear behavior with big edge stresses (Figure 3.5).

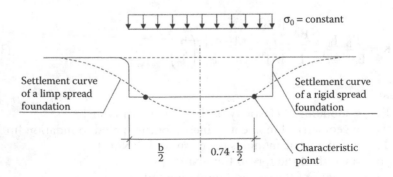

Figure 3.7 Characteristic point of a rectangular spread foundation.

For rigid spread foundations, single foundations, and strip foundations with a big thickness, the distribution of the contact pressure can be determined according to Boussinesq, or with the stress trapeze method [16]. Otherwise, more detailed studies or sufficient conservative assumptions that are "on the safe side" become necessary.

3.3.2.2 Distribution of the contact pressure under rigid foundations according to Boussinesq

Based on the assumption that the subsoil is modeled as an elastic isotropic half-space, in the year 1885, Boussinesq developed equations that can be used for rigid spread foundations in simple cases [17].

The distribution of the contact pressure under a rigid strip foundation with a width b is given by Equation 3.4 (Figure 3.8). For an eccentric load with an eccentricity e, Borowicka enhanced the following equations [22]:

$$\sigma_0 = \frac{2 \cdot V}{\pi \cdot b} \cdot \frac{1}{\sqrt{1-\xi^2}} \quad \text{where} \quad \xi = \frac{2 \cdot x}{b} \tag{3.4}$$

$$e \le \frac{b}{4}, \sigma_0 = \frac{2 \cdot V}{\pi \cdot b} \cdot \frac{1+\left(4 \cdot \frac{e \cdot \xi}{b}\right)}{\sqrt{1-\xi^2}} \tag{3.5}$$

$$e > \frac{b}{4} : \sigma_0 = \frac{2 \cdot V}{\pi \cdot b} \cdot \frac{1+\xi_1}{\sqrt{1-\xi_1^2}} \quad \text{where} \quad \xi_1 = \frac{2x+b-4e}{2b-4e} \tag{3.6}$$

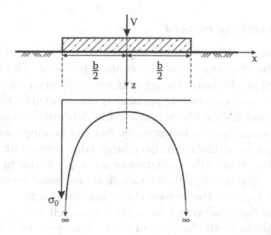

Figure 3.8 Distribution of the contact pressure under rigid foundations according to Boussinesq.

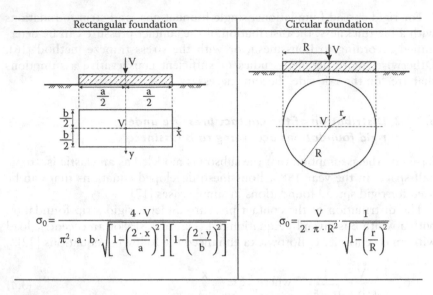

Figure 3.9 Distribution of the contact pressure under rigid foundations as a result of centric loads on elastic isotropic half-space

For circular and rectangular rigid spread foundations, the distribution of the contact pressure can be determined using Figure 3.9.

At the edge of spread foundations, infinitely large stresses arise. Due to the ultimate bearing capacity, imposed by the shear strength of the subsoil, these peak stresses cannot occur. The subsoil plasticizes at the edges of the foundations and the stresses shift to the center of the foundations [23].

3.3.2.3 Stress trapeze method

The stress trapeze method is a statically defined method, and it is the oldest to determine the distribution of the contact pressure. The stress trapeze method is based on the beam theory of elastostatic principles.

The distribution of the contact pressure is determined by the equilibrium conditions ΣV and ΣM, without considering deformations of the building or the subsoil interactions, respectively. Subsoil is simplified with linear elastic behaviour for calculation. Even large edge stresses are theoretically possible. The detection of the reduction of stress peaks due to plasticization is not immediately possible. All considerations are based on the assumption of Bernoulli stating that the cross-sections remain plane.

The force V is the resultant of the applied load, self-weight and buoyancy force. The resultant of the forces and the contact pressures have the same line of influence and the same value but point in opposite directions. To determine the distribution of the contact pressure of an arbitrarily spread

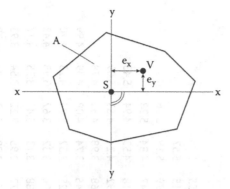

Figure 3.10 Coordinate system for the contact pressure (stress trapeze method).

foundation, Equation 3.7 is used. For the axes of coordinates, an arbitrarily rectangular coordinate system is used, where the zero point corresponds with the center of gravity of the subface (Figure 3.10) [24].

$$\sigma_0 = \frac{V}{A} + \frac{M_y \cdot I_x - M_x \cdot I_{xy}}{I_x \cdot I_y - I_{xy}^2} \cdot x + \frac{M_x \cdot I_y - M_y \cdot I_{xy}}{I_x \cdot I_y - I_{xy}^2} \cdot y \qquad (3.7)$$

If the x- and the y-axis are the main axes of the coordinate system, the centrifugal moment $I_{xy} = 0$. Equation 3.7 is simplified to the following Equation 3.8. If the resultant force V acts at the center of gravity of the subface, the torques $M_x = M_y = 0$. The result is a constant distribution of the contact pressure according to Equation 3.9.

$$\sigma_0 = \frac{V}{A} + \frac{M_y}{I_y} \cdot x + \frac{M_x}{I_x} \cdot y \qquad (3.8)$$

$$\sigma_0 = \frac{V}{A} \qquad (3.9)$$

If the eccentricity of the resulting forces V is too large, theoretically tensile stresses occur, which are not absorbed by the system subsoil-superstructure. An open gap occurs. In this case, Equations 3.7 through 3.9 are not applicable, and the determination of the maximum contact pressure is performed according to the following equation in conjunction with Table 3.2:

$$\sigma_{0,max} = \mu \cdot \frac{V}{A} \qquad (3.10)$$

Table 3.2 Coefficients μ for determining the maximum of the soil contact pressure

	0.00	0.02	0.04	0.06	0.08	0.10	0.12	0.14	0.16	0.18	0.20	0.22	0.24	0.26	0.28	0.30	0.32
0.32	3.70	3.93	4.17	4.43	4.70	4.99											
0.30	3.33	3.54	3.75	3.98	4.23	4.49	4.78	5.09	5.43								
0.28	3.03	3.22	3.41	3.62	3.84	4.08	4.35	4.63	4.94	5.28	5.66						
0.26	2.78	2.95	3.13	3.32	3.52	3.74	3.98	4.24	4.53	4.84	5.19	5.57					
0.24	2.56	2.72	2.88	3.06	3.25	3.46	3.68	3.92	4.18	4.47	4.79	5.15	5.55				
0.22	2.38	2.53	2.68	2.84	3.02	3.20	3.41	3.64	3.88	4.15	4.44	4.77	5.15	5.57			
0.20	2.22	2.36	2.50	2.66	2.82	2.99	3.18	3.39	3.62	3.86	4.14	4.44	4.79	5.19	5.66		
0.18	2.08	2.21	2.35	2.49	2.64	2.80	2.98	3.17	3.38	3.61	3.86	4.15	4.47	4.84	5.28		
0.16	1.96	2.08	2.21	2.34	2.48	2.63	2.80	2.97	3.17	3.38	3.62	3.88	4.18	4.53	4.94	5.43	
0.14	1.84	1.96	2.08	2.21	2.34	2.48	2.63	2.79	2.97	3.17	3.39	3.64	3.92	4.24	4.63	5.09	
0.12	1.72	1.84	1.96	2.08	2.21	2.34	2.48	2.63	2.80	2.98	3.18	3.41	3.68	3.98	4.35	4.78	
0.10	1.60	1.72	1.84	1.96	2.08	2.21	2.34	2.48	2.63	2.80	2.99	3.20	3.46	3.74	4.08	4.49	4.99
0.08	1.48	1.60	1.72	1.84	1.96	2.08	2.21	2.34	2.48	2.64	2.82	3.02	3.25	3.52	3.84	4.23	4.70
0.06	1.36	1.48	1.60	1.72	1.84	1.96	2.08	2.21	2.34	2.49	2.66	2.84	3.06	3.32	3.62	3.98	4.43
0.04	1.24	1.36	1.48	1.60	1.72	1.84	1.96	2.08	2.21	2.35	2.50	2.68	2.88	3.13	3.41	3.75	4.17
0.02	1.12	1.24	1.36	1.48	1.60	1.72	1.84	1.96	2.08	2.21	2.36	2.53	2.72	2.95	3.22	3.54	3.93
0.00	1.00	1.12	1.24	1.36	1.48	1.60	1.72	1.84	1.96	2.08	2.22	2.38	2.56	2.78	3.03	3.33	3.70

e_b/b

3.3.2.4 Subgrade reaction modulus method

Historically, an interaction between soil and structure was taken into account for the first time by using the subgrade reaction modulus method. The prepared subgrade reaction in relation with the change of shape was formulated in the nineteenth century by Winkler [25]. It was created for the design of railway tracks.

According to Winkler, the elastic model of the subsoil, which is also called half-space of Winkler, is a spring model, where at any point the contact pressure σ_0 is proportional to the settlement s (Equation 3.11). The constant of proportionality k_s is called the subgrade reaction modulus. The subgrade reaction modulus can be interpreted as a spring due to the linear mechanical approach for the subsoil behavior (Figure 3.11). However, this method does not consider the interactions between the independent, free-movable vertical springs.

$$\sigma_0(x) = k_s \cdot s(x) \qquad\qquad 3.11$$

where:

σ_0 = contact pressure [kN/m²]
s = settlement [m]
k_s = subgrade reaction modulus [kN/m³]

Using the beam-bending theory, it is possible to describe the curve of the bending moment for an arbitrary, infinitely long and elastic strip foundation with the width b, which is founded on the half-space of Winkler.

The curve of the bending moment of the strip foundation with the bending stiffness $E_b \times I$ is given by

$$M(x) = -E_b \cdot I \cdot \frac{d^2 s(x)}{dx^2} \qquad\qquad (3.12)$$

The double differentiation of Equation 3.12 results in

$$\frac{d^2 M(x)}{dx^2} = -q(x) = -E_B \cdot I \cdot \frac{d^4 s(x)}{dx^4} \qquad\qquad (3.13)$$

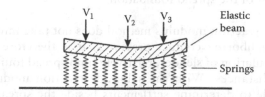

Figure 3.11 Spring model for the subgrade reaction modulus method.

The action q(x) corresponds to the contact pressure $\sigma_0(x)$, which can be described by

$$q(x) = -\sigma_0(x) \cdot b = -k_s \cdot s(x) \cdot b = E_B \cdot I \cdot \frac{d^4 s(x)}{dx^4} \qquad (3.14)$$

With the elastic length L given as

$$L = \sqrt[4]{\frac{4 \cdot E_B \cdot I}{k_s \cdot b}} \qquad (3.15)$$

and the elimination of s(x), Equation 3.16 follows. For a large number of boundary conditions, Equation 3.16 can be solved. For an infinite long strip foundation, the distribution of the contact pressure $\sigma_0(x)$, the distribution of the bending moment M(x), and the distribution of the shear forces result according to Equations 3.17 through 3.19.

$$\frac{d^4 M(x)}{dx^4} + \frac{4\, M(x)}{L^4} = 0 \qquad (3.16)$$

$$\sigma_0 = \frac{V}{2 \cdot b \cdot L} \cdot e^{-\frac{x}{L}} \cdot \left(\cos\frac{x}{L} + \sin\frac{x}{L} \right) \qquad (3.17)$$

$$M(x) = \frac{V \cdot L}{4} \cdot e^{-\frac{x}{L}} \cdot \left(\cos\frac{x}{L} - \sin\frac{x}{L} \right) \qquad (3.18)$$

$$Q(x) = \pm\frac{V}{2} \cdot e^{-\frac{x}{L}} \cdot \cos\frac{x}{L} \qquad (3.19)$$

The subgrade reaction modulus is not a soil parameter. It depends on the following parameters:

- Oedometric modulus of the subsoil
- Thickness of the compressible layer
- Dimensions of the spread foundation

The subgrade reaction modulus method does not take into account the influence of neighboring contact pressures. It is therefore mainly suitable for the calculation of slender, relatively limp spread foundations with large column distances. With the subgrade reaction modulus method, it is not possible to determine settlements beside the spread foundation (Figure 3.12).

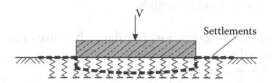

Figure 3.12 Distribution of the settlements according to the subgrade reaction modulus method.

3.3.2.5 *Stiffness modulus method*

The stiffness modulus method according to Ohde (1942) describes the soil–structure interaction more accurately than the subgrade reaction modulus method, because the influence of adjacent contact pressures is considered on the settlement of an arbitrary point of the spread foundation [19,26]. In the stiffness modulus method, the bending moment of the linear elastic modeled spread foundation is linked with the bending moment of the linear elastic, isotropic modeled settlement trough. The same deformations arise.

Figure 3.13 represents the distribution of the settlement of a spread foundation according to the stiffness modulus method.

In geotechnical engineering practice, spread foundations with complex load situations and geometric boundary conditions are normally examined using computer programs. For most cases, no closed solutions are available for the statically indeterminate system of equations.

The assumption of infinite elastic subsoil has the consequence that theoretically infinite large stress peaks result at the edge of the spread foundation. Due to the plasticizing effect of the subsoil, these stress peaks do not occur in reality. Powerful computer programs consider this basic soil mechanical behavior.

3.3.3 Geotechnical analysis

In the following section the geotechnical analysis for stability and serviceability of spread foundations are defined according to the currently valid technical regulations EC 7.

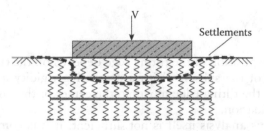

Figure 3.13 Distribution of the settlements according to the stiffness modulus method.

The analysis of the stability includes

- Analysis of safety against loss of balance because of overturning
- Analysis of safety against sliding
- Analysis of safety against base failure
- Analysis of safety against buoyancy

The analysis of the serviceability includes

- Analysis of the foundation rotation and the limitation of the open gap
- Analysis of horizontal displacements
- Analysis of settlements and differential settlements

3.3.3.1 Analysis of safety against loss of balance because of overturning

Up to now, the analysis of safety against loss of balance because of over-turning was done by applying the resultant of the forces into the second core width. That means that the lower surface of the spread foundation has only a small part with an open gap. This approach is described in [27,28]. Thus, a resulting force in the first core width creates a compressive stress over the entire lower surface of the spread foundation.

According to the current technical regulations, the analysis of safety against loss of balance because of overturning is based on a principle of the rigid body mechanics. The destabilizing and stabilizing forces are compared based on a fictional tilting edge at the edge of the spread foundation:

$$E_{dst,d} \leq E_{stb,d} \tag{3.20}$$

The design value of the destabilizing force is estimated according to Equation 3.21, and the design value of the stabilizing action is estimated according to Equation 3.22:

$$E_{dst,d} = E_{G,dst,k} \cdot \gamma_{G,dst} + E_{Q,dst,k} \cdot \gamma_{Q,dst} \tag{3.21}$$

$$E_{stb,d} = E_{stb,k} \cdot \gamma_{G,stb} \tag{3.22}$$

In reality, the position of the tilting edge depends on the rigidity and the shear strength of the subsoil. With a decreasing rigidity and decreasing shear strength, the tilting edge moves to the center of the lower surface of the spread foundation.

Therefore, this analysis itself is not sufficient. It was complemented by the analysis of the limitation of the open gap, which is defined for the serviceability limit state. According to [10], the resultant force of the permanent loads has to be applied into the first core width and the resultant force of the variable loads has to be applied into the second core width (Figure 3.21).

3.3.3.2 Analysis of safety against sliding

The analysis of safety against sliding (limit state GEO-2) is calculated according to Equation 3.23. The forces parallel to the lower surface of the spread foundation have to be smaller than the total resistance, consisting of slide resistance and passive earth pressure. If the passive earth pressure is considered, the serviceability limit state has to be verified regarding the horizontal displacements.

$$H_d \leq R_d + R_{p,d} \tag{3.23}$$

where: $R_d = \dfrac{R_k}{\gamma_{R,h}}$

$R_{p,d} = \dfrac{R_{p,k}}{\gamma_{R,h}}$

The sliding resistance is determined according to the three following cases:

- Sliding in the gap between the spread foundation and in the subjacent, fully consolidated subsoil:

$$R_d = \dfrac{V_k \cdot \tan\delta}{\gamma_{R,h}} \tag{3.24}$$

where:
V_k = characteristic value of the vertical loadings [kN]
δ = characteristic value of the angle of base friction [°]

- Sliding when the gap passes through the fully consolidated soil, for example, in the arrangement of a foundation cut-off:

$$R_d = \dfrac{V_k \cdot \tan\varphi' + A \cdot c'}{\gamma_{R,h}} \tag{3.25}$$

where:
V_k = characteristic value of the vertical loadings [kN]
φ' = characteristic friction angle for the subsoil under the spread foundation [°]

A = area of the load transmission [m²]

c' = characteristic value of the cohesion of the subsoil [kN/m²]

- Sliding on water-saturated subsoil due to very quick loading:

$$R_d = \frac{A \cdot c_u}{\gamma_{R,h}}$$ (3.26)

where:

A = Area of the load transmission [m²]

c_u = Characteristic value of the undrained cohesion of the subsoil [kN/m²]

For spread foundations that are concreted *in situ*, the characteristic value of the angle of base friction δ is the same as the characteristic value of the friction angle φ' of the soil. For prefabricated spread foundation elements the characteristic value of the angle of base friction δ should be set to 2/3 φ'. The characteristic value of the angle of base friction should be $\delta \leq 35°$.

A passive earth pressure can be considered if the spread foundation is deep enough. Due to horizontal deformations, the passive earth pressure should be limited to 50% of the possible passive earth pressure. Basically, it must be verified whether the passive earth pressure exists during all possible stages in the construction and the service phase of the foundation.

3.3.3.3 Analysis of safety against base failure

The analysis of safety against base failure is guaranteed if the design value of the bearing capacity R_d is bigger than the design value of the active force V_d. R_d is calculated according to Equation 3.27. The principle scheme of a bearing failure of a spread foundation is shown in Figure 3.14.

$$R_d = \frac{R_{n,k}}{\gamma_{R,v}}$$ (3.27)

The resistance of the bearing capacity is determined by the soil properties (density, shear parameters), the dimensions and the embedment depth of the spread foundation. Detailed information can be found in the incidental standard [29,30]. The characteristic resistance of the bearing capacity $R_{n,k}$ is calculated analytically with a trinomial equation, which is based on the moment equilibrium of the failure figure of the bearing capacity in ideal plastic, plane deformation state [31]. The trinomial equation of the bearing capacity considers the width b of the foundation, the embedment depth d of the foundation and the cohesion c' of the subsoil. All three aspects have to be factorized with the bearing capacity factors N_b, N_d, and N_c:

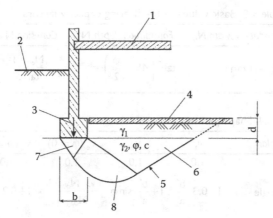

Figure 3.14 Bearing failure figure of a strip foundation 1, Reinforced wall; 2, area; 3, resulting contact pressure; 4, cellar floor; 5, sliding surface, the form depends on the angle of friction φ; 6, passive Rankine-zone of the failure body; 7, active Rankine-zone of the failure body; 8, Prandtl-zone of the failure body.

$$R_{n,k} = a' \cdot b' \cdot \left(\gamma_2 \cdot b' \cdot N_b + \gamma_1 \cdot d \cdot N_d + c' \cdot N_c \right) \qquad (3.28)$$

where:

$$N_b = N_{b0} \cdot v_b \cdot i_b \cdot \lambda_b \cdot \xi_b$$

$$N_d = N_{d0} \cdot v_d \cdot i_d \cdot \lambda_d \cdot \xi_d$$

$$N_c = N_{c0} \cdot v_c \cdot i_c \cdot \lambda_c \cdot \xi_c$$

The density γ_1 describes the density of the subsoil above the foundation level. The density γ_2 describes the density of the subsoil underneath the foundation level. The bearing capacity factors N_b, N_d, and N_c consider the following boundary conditions:

- Basic values of the bearing capacity factors: N_{b0}, N_{d0}, N_{c0}
- Shape parameters: v_b, v_d, v_c
- Parameter for load inclination: i_b, i_d, i_c
- Parameters for landscape inclination: λ_b, λ_d, λ_c
- Parameters for the base inclination: ξ_b, ξ_d, ξ_c

The parameters of the bearing capacity factors N_{b0}, N_{d0}, N_{c0} depend on the angle of friction of the subsoil φ' and are calculated according to Table 3.3.

Table 3.3 Basic values of the bearing capacity factors

Foundation width N_{b0}	Foundation depth N_{d0}	Cohesion N_{c0}
$(N_{d0}-1)\tan\varphi$	$\tan^2\left(45° + \dfrac{\varphi}{2}\right)\cdot e^{\pi\cdot\tan\varphi}$	$\dfrac{N_{d0}-1}{\tan\varphi}$

Table 3.4 Shape parameters v_i

Floor plan	v_b	v_d	$v_c\ (\varphi \neq 0)$	$v_c\ (\varphi = 0)$
Strip	1.0	1.0	1.0	1.0
Rectangle	$1 - 0.3\cdot\dfrac{b'}{a'}$	$1 + \dfrac{b'}{a'}\cdot\sin\varphi$	$\dfrac{v_d\cdot N_{d0}-1}{N_{d0}-1}$	$1 + 0.2\cdot\dfrac{b'}{a'}$
Square/Circle	0.7	$1 + \sin\varphi$	$\dfrac{v_d\cdot N_{d0}-1}{N_{d0}-1}$	1.2

The shape parameters v_b, v_d, v_c take into account the geometric dimensions of the spread foundation. For the standard applicable geometry, the shape parameters are summarized in Table 3.4.

If eccentric forces have to be considered the base area has to be reduced. The resulting load has to be in the center of gravity. The reduced dimensions a' and b' are calculated according to Equations 3.29 and 3.30. Basically applied, is $a > b$ and $a' > b'$, respectively. For spread foundations with open parts, the external dimensions may be used for the analysis, if the open parts are not bigger than 20% of the whole base area.

$$a' = a - 2e_a \tag{3.29}$$

$$b' = b - 2e_b \tag{3.30}$$

$$m = m_a \cdot \cos^2\omega + m_b \cdot \sin^2\omega \tag{3.31}$$

where $m_a = \dfrac{2 + \dfrac{a'}{b'}}{1 + \dfrac{a'}{b'}}$ and $m_b = \dfrac{2 + \dfrac{b'}{a'}}{1 + \dfrac{b'}{a'}}$

Forces T_k that are parallel to the foundation level are considered by the parameters i_b, i_d, i_c for the load inclination. The definition of the angle of the load inclination is shown in Figure 3.15. The determination of the parameters for the load inclination is shown in Tables 3.5 and 3.6. The orientation of the acting forces is determined by the angle ω (Figure 3.16). For a strip foundation $\omega = 90°$.

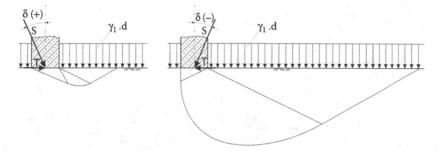

Figure 3.15 Definition of the angle of the load inclination.

Table 3.5 Parameter i_i for load inclination if $\varphi' > 0$

Direction	i_b	i_d	i_c
$\delta > 0$	$(1 - \tan \delta)^{m+1}$	$(1 - \tan \delta)^m$	$\dfrac{i_d \cdot N_{d0} - 1}{N_{d0} - 1}$
$\delta < 0$	$\cos \delta \cdot (1 - 0.04 \cdot \delta)^{0.64 + 0.028 \cdot \varphi}$	$\cos \delta \cdot (1 - 0.0244 \cdot \delta)^{0.03 + 0.04 \varphi}$	

Table 3.6 Parameter i_i of the load inclination if $\varphi' = 0$

i_b	i_d	i_c
Not necessary, because of $\varphi = 0$	1.0	$0.5 + 0.5\sqrt{1 - \dfrac{T_k}{A' \cdot c}}$

An inclination of the landscape is considered by the parameters λ_b, λ_d, λ_c for landscape inclination. The parameters depend on the slope inclination β. The slope inclination has to be less than the angle of friction φ' of the subsoil and the longitudinal axis of the foundation has to be parallel to the slope edge. The determination of the parameters for landscape inclination is shown in Figure 3.17 and Table 3.7.

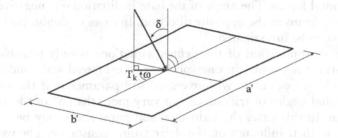

Figure 3.16 Angle ω for an oblique acting load.

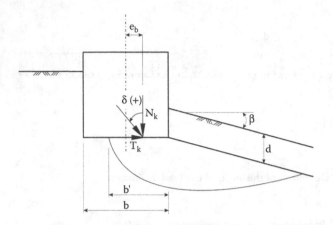

Figure 3.17 Eccentric, oblique loaded strip foundation on a slope.

Table 3.7 Parameters λ_i for landscape inclination

Case	λ_b	λ_d	λ_c
$\varphi > 0$	$(1 - 0.5 \tan\beta)^6$	$(1 - \tan\beta)^{1.9}$	$\dfrac{N_{d0} \cdot e^{-0.0349 \cdot \beta \cdot \tan \varphi} - 1}{N_{d0} - 1}$
$\varphi = 0$	—	1.0	$1 - 0.4 \tan\beta$

Table 3.8 Coefficient ξ_i of the base inclination

Case	ξ_b	ξ_d	ξ_c
$\varphi > 0$	$e^{-0.045 \cdot \alpha \cdot \tan \varphi}$	$e^{-0.045 \cdot \alpha \cdot \tan \varphi}$	$e^{-0.045 \cdot \alpha \cdot \tan \varphi}$
$\varphi = 0$	—	1.0	$1 - 0.0068\alpha$

The base inclination is considered by the parameters ξ_b, ξ_d, ξ_c for the base inclination (Table 3.8), which depend on the angle of friction φ' of the subsoil and the base inclination α of the spread foundation. The definition of the base inclination is shown in Figure 3.18. The angle of the base inclination α is positive, if the failure body forms out in the direction of the horizontal forces. The angle of the base inclination α is negative, if the failure body forms in the opposite direction. In cases of doubt, both failure bodies have to be investigated.

The direct application of the defined equations is only possible if the sliding surface is formed in one soil layer. For layered soil conditions, it is permissible to calculate with averaged soil parameters if the values of the individual angles of friction do not vary more than 5° of the arithmetic average. In this case, the individual soil parameters may be weighted according to their influence on the shear failure resistance. The weighting is as follows.

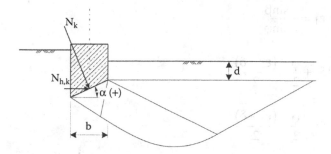

Figure 3.18 Angle of the base inclination α.

- The average density is related to the percentage of the individual layers in the cross-section area of the failure body
- The average angle of friction and the average cohesion are related to the percentage of the individual layers in the cross-section area of the failure body

Authoritative for the sliding surface is the average of the angle of friction φ. To determine whether the failure body has more than one layer, it is recommended to define the failure body according to Equation 3.32 through 3.38 (Figure 3.19). For simple cases (α = β = δ = 0), the Equations 3.39 through 3.42 have to be applied.

$$\vartheta_1 = 45° - \frac{\varphi}{2} - \frac{(\varepsilon_1 + \beta)}{2} \tag{3.32}$$

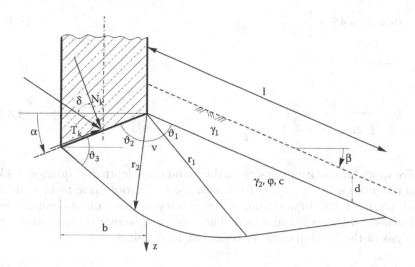

Figure 3.19 Determination of the failure body.

where: $\sin \varepsilon_1 = -\dfrac{\sin \beta}{\sin \varphi}$

$$\vartheta_2 = 45° + \frac{\varphi}{2} - \frac{(\varepsilon_2 - \delta)}{2} \tag{3.33}$$

$$\vartheta_3 = 45° + \frac{\varphi}{2} + \frac{(\varepsilon_2 - \delta)}{2} \tag{3.34}$$

where $\sin \varepsilon_2 = -\dfrac{\sin \delta}{\sin \varphi}$

$$\nu = 180° - \alpha - \beta - \vartheta_1 - \vartheta_2 \tag{3.35}$$

$$r_2 = b' \cdot \frac{\sin \vartheta_3}{\cos \alpha \cdot \sin(\vartheta_2 + \vartheta_3)} \tag{3.36}$$

$$r_1 = r_2 \cdot e^{0.0175 \cdot \nu \cdot \tan \varphi} \tag{3.37}$$

$$l = r_1 \cdot \frac{\cos \varphi}{\cos(\vartheta_1 + \varphi)} \tag{3.38}$$

$$\vartheta_1 = 45° - \frac{\varphi}{2} \tag{3.39}$$

$$\vartheta_2 = \vartheta_3 = 45° + \frac{\varphi}{2} \tag{3.40}$$

$$\nu = 90° \tag{3.41}$$

$$r_2 = \frac{b'}{2 \cdot \cos\left(45° + \dfrac{\varphi}{2}\right)} \tag{3.42}$$

For spread foundations at slopes, the foundation depth d′ (Equation 3.43) and the parameters λ_b, λ_d, λ_c for landscape inclination have to be considered (Figure 3.20). In addition, it is necessary to carry out a comparative calculation with β = 0 and d′ = d. The smaller resistance is the basis of the analysis of the bearing capacity regarding base failure.

$$d' = d + 0.8 \cdot s \cdot \tan \beta \tag{3.43}$$

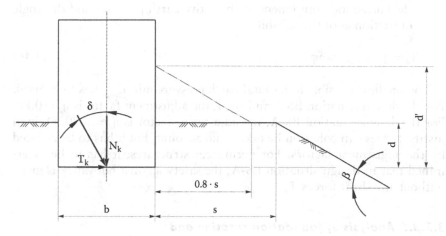

Figure 3.20 Spread foundation on a slope.

3.3.3.4 Analysis of safety against buoyancy

The analysis of safety against buoyancy (limit state UPL) is performed using Equation 3.44. This equation is the proof that the net weight of the structure is big enough compared to the buoyant force of the water. Shear forces (friction forces at the side) can only be considered if the transmission of the forces is ensured. Acting shear forces T_k may be

$$G_{dst,k} \cdot \gamma_{G,dst} + Q_{dst,rep} \cdot \gamma_{Q,dst} \le G_{stb,k} \cdot \gamma_{G,stb} + T_k \cdot \gamma_{G,stb} \qquad (3.44)$$

where:

$G_{dst,k}$	= permanent destabilizing vertical load (buoyancy)
$\gamma_{G,dst}$	= partial safety factor for permanent destabilizing load
$Q_{dst,rep}$	= representative variable destabilizing vertical load
$\gamma_{Q,dst}$	= partial safety factor for variable destabilizing load
$G_{stb,k}$	= permanent stabilizing load
$\gamma_{G,stb}$	= partial safety factor for permanent stabilizing load
T_k	= shear force

- Vertical component of the active earth pressure $E_{av,,k}$ on a retaining structure depending on the horizontal component of the active earth pressure $E_{ah,k}$ as well as the angle of wall friction δ_a (Equation 3.45)

$$T_k = \eta_z \cdot E_{ah,k} \cdot \tan\delta_a \qquad (3.45)$$

- Vertical component of the active earth pressure in a joint of the subsoil, for example, starting at the end of a horizontal spur in dependency to the horizontal component of the active earth pressure and the angle of friction φ' of the subsoil:

$$T_k = \eta_z \cdot E_{ah,k} \cdot \tan\varphi' \tag{3.46}$$

The smallest possible horizontal earth pressure min $E_{ah,k}$ has to be used. For the design situation BS-P and BS-T, the adjustment factor is $\eta_z = 0.80$. For the design situation BS-A, the adjustment factor is $\eta_z = 0.90$. Only in justified cases can cohesion be taken into account, but it has to be reduced by the adjustment factors. For permanent structures, it has to be determined that in design situation BS-A, the safety against buoyancy is given without any shear forces T_k.

3.3.3.5 Analysis of foundation rotation and limitation of the open gap

Generally, the serviceability limit states refer to absolute deformations and displacements as well as differential deformations. In special cases, for example, time-dependent material behavior displacement rates have to be considered.

For the analysis of foundation rotation and limitation of the open gap, the resultant of the dead loads has to be limited into the first core width, which means that an open gap does not occur. The first core width for rectangular spread foundations can be determined according to Equation 3.47. For circular spread foundations Equation 3.48 is used. Furthermore, it should be guaranteed that the resultant of the permanent loads and the variable loads are in the second core width, so an open gap cannot occur across the center line of the spread foundation. The second core width for rectangular layouts can be determined according to Equation 3.49. For circular spread foundations Equation 3.50 is used. Figure 3.21 shows the first and the second core width for a rectangular spread foundation.

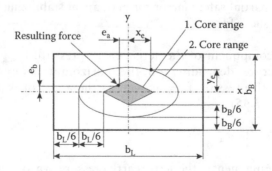

Figure 3.21 Limitation of the eccentricity.

$$\frac{x_e}{a} + \frac{y_e}{b} = \frac{1}{6} \tag{3.47}$$

$$e \leq 0.25 \cdot r \tag{3.48}$$

$$\left(\frac{x_e}{a}\right)^2 + \left(\frac{y_e}{b}\right)^2 = \frac{1}{9} \tag{3.49}$$

$$e \leq 0.59 \cdot r \tag{3.50}$$

For single and strip foundations, which are founded on medium-dense, non-cohesive soils and stiff cohesive soils, respectively, no incompatible distortions of the foundation can be expected if the acceptable eccentricity is observed.

The analysis of foundation rotation and limitation of the open gap is mandatory according to [10], if the analysis of safety against loss of balance because of overturning is carried out by using one edge of the spread foundation as a tilting edge.

3.3.3.6 Analysis of horizontal displacements

Generally, for spread foundations the analysis of horizontal displacement is observed, if:

- The analysis of safety against sliding is performed without considering a passive earth pressure.
- For medium-dense, non-cohesive soils and stiff cohesive soils, respectively, only two-thirds of the characteristic sliding resistance in the foundation level and not more than one-third of the characteristic earth pressure are considered.

If these arguments are not true, it is necessary to analyze the possible horizontal displacements. Permanent loads and variable loads, as well as infrequent or unique loads, have to be considered.

3.3.3.7 Analysis of settlements and differential settlements

The determinations for settlements of spread foundations are conducted in accordance with [32]. Normally, the influence depth of the contact pressure is between $z = b$ and $z = 2b$.

Due to the complex interaction between the subsoil and the construction, it is difficult to provide information about the acceptable settlements or differential settlements for constructions [33]. Figure 3.22 indicates the damage factors for the angular distortion as a result of settlements [33–35].

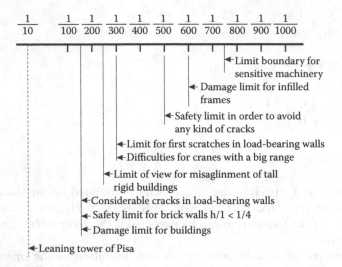

Figure 3.22 Damage criterion.

Regarding the tilting of high-rise structures, the analysis of safety against inclination has to check that the occurring tilting is harmless for the structure [33]. The analysis for rectangular spread foundations is performed according to Equation 3.51. The analysis for circular spread foundations is performed according to Equation 3.52.

$$\frac{b^3 \cdot E_m}{V_d \cdot h_s \cdot f_y} \geq 1 \tag{3.51}$$

$$\frac{r^3 \cdot E_m}{V_d \cdot h_s \cdot f_y} \geq 1 \tag{3.52}$$

In Equations 3.51 and 3.52:
E_m = Modulus for the compressibility of the subsoil
h_s = Height of the center of gravity above the foundation level
f_r and f_y = Tilting coefficients
V_d = Design value of the vertical loads
More detailed information can be found in [33] and [36].

3.3.3.8 Simplified analysis of spread foundations in standard cases

The simplified analysis of spread foundations in standard cases consists of a simple comparison between the base resistance $\sigma_{R,d}$ and the contact pressure $\sigma_{E,d}$ (Equation 3.53). For spread foundations with the area $A = a \times b$

or $A' = a' \times b'$, the analysis of safety against sliding and base failure, as well as the analysis of the serviceability limit state, can be applied in standard cases. These standard cases include:

- Horizontal lower surface of the foundation and an almost horizontal landscape and soil layers
- Sufficient strength of the subsoil into a depth of twice the width of the foundation below the foundation level (minimum 2 m)
- No regular dynamic or mainly dynamic loads; no pore water pressure in cohesive soils
- Passive earth pressure can only be applied if it is assured by constructive or other procedures
- The inclination of the resultant of the contact pressure observes the rule $\tan\delta = H_k/V_k \leq 0.2$ (δ = inclination of the resultant of the contact pressure; H_k = characteristic horizontal forces; V_k = characteristic vertical forces)
- The admissible eccentricity of the resultant of the contact pressure is observed
- The analysis of safety against loss of balance because of overturning is observed

$$\sigma_{E,d} \leq \sigma_{R,d} \tag{3.53}$$

The design values of the contact pressure $\sigma_{R,d}$ are based on the combined examination of the base failure and the settlements. If only the SLS is analyzed, the admissible contact pressure increases with an increasing width of the spread foundation. If only the ULS is analyzed, the admissible contact pressure decreases with an increasing width of the spread foundation. Figure 3.23 shows the two fundamental demands for an adequate analysis against base failure (ULS) and the analysis of the settlements (SLS). For foundation widths that are bigger than the width b_s, the acceptable contact pressure decreases because of settlements.

The design values of the contact pressure $\sigma_{R,d}$ for simplified analysis of strip foundations are specified in tables. The tabular values can also be used for single foundations [10,37,38].

If the foundation level is more than 2 m below the surface on all sides, the tabular values can be raised. The raise can be 1.4 times the unloading due to the excavation below a depth of ≥ 2 m under the surface.

The settlement values in the tables refer to detached strip foundations with a central loading (no eccentricity). If eccentric loads occur, the serviceability has to be analyzed. For the application of the current table values, it is essential to notice that in earlier editions of these tables, characteristic values were given [10].

The simplified analysis of the ULS and SLS of strip foundations in non-cohesive soils considers the design situation BS-P. For the design situation

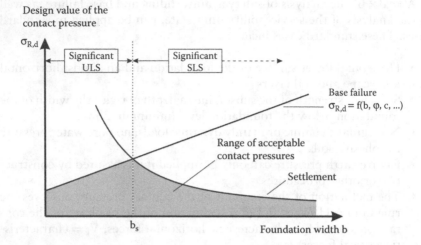

Figure 3.23 Maximum contact pressure $\sigma_{R,d}$ taking into account the stability (ULS) and serviceability (SLS).

Table 3.9 Requirements for the application of the design values $\sigma_{R,d}$ in non-cohesive soils

Soil group according to DIN 18196	Coefficient of uniformity according to DIN 18196 C_u	Compactness according to DIN 18126 D	Compression ratio according to DIN 18127 D_{Pr}	Point resistance of the soil penetrometer q_c [MN/m²]
SE, GE, SU, GU, ST, GT	≤ 3	≥ 0.30	≥ 95%	≥ 7.5
SE, SW, SI, GE GW, GT, SU, GU	> 3	≥ 0.45	≥ 98%	≥ 7.5

BS-T the tabular values are "on the safe side". The tabular values are applicable for vertical loads. Intermediate values may be interpolated linearly. For eccentric loads the tabular values may be extrapolated if width b′ < 0.50 m. There must be a distance between the lower surface of the foundation and the groundwater level. The distance must be bigger than the width b or b′ of the foundation. For the application of the tables for non-cohesive soils the requirements in Table 3.9 must be fulfilled. The short forms of the soil groups are explained in Table 3.10.

The coefficient of uniformity C_u describes the gradient of the grain size distribution in the area of passing fractions of 10% and 60% and it is determined according to Equation 3.54 [39]. According to [40], the compactness D describes whether the subsoil is loose, medium dense, or dense. The compactness D is determined by the porosity n, according to Equation 3.55. The compression ratio D_{pr} is the relation between the

Table 3.10 Explanation of the soil groups

Short form according to DIN 18196	Long form according to DIN 18196 in German	Long form according to DIN 18196 in English
SE	Sand, enggestuft	Sand with small grain size distribution
SW	Sand, weitgestuft	Sand with a wide spreaded grain size distribution
SI	Sand, intermittierend gestuft	Sand with an intermittent spreaded grain size distribution
GE	Kies, enggestuft	Gravel with small grain size distribution
GW	Kies, weitgestuft	Gravel with a wide spreaded grain size distribution
ST	Sand, tonig (Feinkornanteil: 5–15%)	Sand, clayey (fine fraction: 5–15%)
SU	Sand, schluffig (Feinkornanteil: 5–15%)	Sand, silty (fine fraction: 5–15%)
GT	Kies, tonig (Feinkornanteil: 5–15%)	Gravel, clayey (fine fraction: 5–15%)
GU	Kies, schluffig (Feinkornanteil: 5–15%)	Gravel, clayey (fine fraction: 5–15%)

proctor density ρ_{pr} (density at optimal water content) and the dry density ρ_d [41]. The compression ratio is calculated using Equation 3.56.

$$C_u = \frac{d_{60}}{d_{10}}$$ (3.54)

$$D = \frac{\max n - n}{\max n - \min n}$$ (3.55)

$$D_{pr} = \frac{\rho_d}{\rho_{pr}}$$ (3.56)

For simplified analysis of strip foundations Table 3.11 shows the permissible design values of the contact pressure $\sigma_{R,d}$ for non-cohesive soils taking into account an adequate safety against base failure. If the settlement has to be limited additionally, Table 3.12 has to be applied. For the purpose of Table 3.12, the settlements are limited to 1–2 cm.

The permissible design values of the contact pressure $\sigma_{R,d}$ for strip foundations in non-cohesive soils with the minimum width $b \geq 0.50$ m and the minimum embedment depth $d \geq 0.50$ m can be increased as follows:

- Increase of the design values by 20% in Table 3.11 and 3.12, if single foundations have an aspect ratio $a/b < 2$ resp. $a'/b' < 2$; for Table 3.11

Table 3.11 Design values $\sigma_{R,d}$ for strip foundations in non-cohesive soils and sufficient safety against hydraulic failure with a vertical resultant of the contact pressure

Smallest embedment depth of the foundation [m]	Design value of the contact pressure $\sigma_{R,d}$ [kN/m²] in dependence of the foundation width b resp. b'					
	0.50 m	1.00 m	1.50 m	2.00 m	2.50 m	3.00 m
0.50	280	420	560	700	700	700
1.00	380	520	660	800	800	800
1.50	480	620	760	900	900	900
2.00	560	700	840	980	980	980
For buildings with an embedment depth 0.30 m $\leq$ d $\leq$ 0.50 m and foundation width b resp. b' $\geq$ 0.30 m			210			

Table 3.12 Design values $\sigma_{R,d}$ for strip foundations in non-cohesive soils and limitation of the settlements to 1–2 cm with a vertical resultant of the contact pressure

Smallest embedment depth of the foundation [m]	Design value of the contact pressure $\sigma_{R,d}$ [kN/m²] in dependence of the foundation width b resp. b'					
	0.50 m	1.00 m	1.50 m	2.00 m	2.50 m	3.00 m
0.50	280	420	460	390	350	310
1.00	380	520	500	430	380	340
1.50	480	620	550	480	410	360
2.00	560	700	590	500	430	390
For buildings with an embedment depth 0.30 m $\leq$ d $\leq$ 0.50 m and foundation width b resp. b' $\geq$ 0.30 m			210			

it is only applied if the embedment depth d is bigger than $0.60 \times b$ resp. $0.60 \times b'$

- Increase of the design values by 50% in Tables 3.11 and 3.12, if the subsoil complies the values in Table 3.13 into a depth of twice the width under the foundation level (minimum 2 m under foundation level)

The permissible design values of the contact pressure for strip foundations in non-cohesive soils in Table 3.11 (even increased and/or reduced due to horizontal loads) have to be reduced if groundwater has to be considered:

- Reduction of the design values by 40%, if the groundwater level is the same as the foundation level

Table 3.13 Requirements for increasing the design values $\sigma_{R,d}$ for non-cohesive soils

Soil group according to DIN 18196	Coefficient of uniformity according to DIN 18196 C_u	Compactness according to DIN 18126 D	Compression ratio according to DIN 18127 D_{Pr}	Point resistance of the soil penetrometer q_c [MN/m²]
SE, GE, SU, GU, ST, GT	≤3	≥0.50	≥98%	≥15
SE, SW, SI, GE GW, GT, SU, GU	>3	≥0.65	≥100%	≥15

- If the distance between the groundwater level and the foundation level is smaller than b or b′, it has to be interpolated between the reduced and the non-reduced design values $\sigma_{R,d}$
- Reduction of the design values by 40%, if the groundwater level is above the foundation level, provided that the embedment depth d ≥ 0.80 m and d ≥ b; a separate analysis is only necessary if both conditions are not true

The permissible design values of the contract pressure $\sigma_{R,d}$ in Table 3.12 can only be used if the design values in Table 3.11 (even increased and/or reduced due to horizontal loads and/or groundwater) are bigger.

The permissible design values of the contact pressure $\sigma_{R,d}$ for strip foundations in non-cohesive soils shown in Table 3.11 (even increased and/or reduced due to groundwater) need to be reduced for a combination of characteristic vertical (V_k) and horizontal (H_k) loads as follows:

- Reduction by the factor $(1 - H_k/V_k)$ if H_k is parallel to the long side of the foundation and if the aspect ratio is a/b ≥ 2 resp. a′/b′ ≥ 2
- Reduction by the factor $(1 - H_k/V_k)^2$ in all other cases

The design values of the contact pressure shown in Table 3.12 can only be applied if the design values $\sigma_{R,d}$ shown in Table 3.11 (even increased and/or reduced due to groundwater) are bigger.

The simplified analysis of the ULS and SLS of strip foundations in cohesive soils is for the design situation BS-P. For the design situation BS-T, the tabular values are "on the safe side." The tabular values are applicable for vertical and inclined loads. Intermediate values may be interpolated linearly. The tables are given for different types of soil. The short forms of the soil groups are explained in Table 3.10. If the Tables 3.14 through 3.17 are used, settlements of 2–4 cm can be expected. In principle, the Tables 3.14 through 3.17 are only applicable for soil types with a granular structure that may not collapse suddenly.

The design values $\sigma_{R,d}$ for strip foundations in cohesive soil shown in Tables 3.14 through 3.17 (even reduced due to foundation width b > 2 m) may be increased by 20% if the aspect ratio is a/b < 2 resp. a′/b′ < 2.

Table 3.14 Design values $\sigma_{R,d}$ for strip foundations in silt

Silt (UL according to DIN 18126) consistency: Solid to semisolid	
Smallest embedment depth of the foundation [m]	Design value $\sigma_{R,d}$ of the contact pressure [kN/m²]
0.50	180
1.00	250
1.50	310
2.00	350
Unconfined compressive strength $q_{u,k}$ [kN/m²]	120

Table 3.15 Design values $\sigma_{R,d}$ for strip foundations in mixed soils

Mixed soils (SU*, ST, ST*, GU*, GT* according to DIN 18196)			
	Design values $\sigma_{R,d}$ of the contact pressure [kN/m²]		
	Consistency		
Smallest embedment depth of the foundation [m]	Stiff	Semi-solid	Solid
0.50	210	310	460
1.00	250	390	530
1.50	310	460	620
2.00	350	520	700
Unconfined compressive strength $q_{u,k}$ [kN/m²]	120–300	300–700	>700

Table 3.16 Design values $\sigma_{R,d}$ for strip foundations in clay, silty soils

Clay, silty soils (UM, TL, TM according to DIN 18196)			
	Design values $\sigma_{R,d}$ of the contact pressure [kN/m²]		
	Consistency		
Smallest embedment depth of the foundation [m]	Stiff	Semi-solid	Solid
0.50	170	240	390
1.00	200	290	450
1.50	220	350	500
2.00	250	390	560
Unconfined compressive strength $q_{u,k}$ [kN/m²]	120–300	300–700	>700

Table 3.17 Design values $\sigma_{R,d}$ for strip foundations in clay

Clay (TA according to DIN 18196)			
	Design values $\sigma_{R,d}$ of the contact pressure [kN/m²]		
	Consistency		
Smallest embedment depth of the foundation [m]	Stiff	Semi-solid	Solid
0.50	130	200	280
1.00	150	250	340
1.50	180	290	380
2.00	210	320	420
Unconfined compressive strength $q_{u,k}$ [kN/m²]	120–300	300–700	>700

The design values $\sigma_{R,d}$ for strip foundations in cohesive soil shown in Tables 3.14 through 3.17 (even increased due to the aspect ratio) have to be reduced by 10% per meter at foundation width b = 2–5 m. For foundations with a width b > 5 m the ULS and the SLS have to be checked separately according to the classic soil mechanical analysis.

3.4 EXAMPLES OF SPREAD FOUNDATIONS FROM ENGINEERING PRACTICE

In the last decades, increasing population density worldwide has led to the construction of more and higher high-rise buildings. Until 1960, in Frankfurt am Main, Germany, buildings with 10–15 storeys were considered to be high-rise. In 1961, the first building with 20 storeys was constructed, and in 1969 the first Commerzbank Tower with 30 storeys and a height of 130 m was completed. In the 1970s and early 1980s, several other skyscrapers were built with heights of 150–180 m. All were founded in the very settlement-active Frankfurt Clay. The experiences in Frankfurt am Main show, that the final settlements of a spread foundation can be 1.7 to 2.0 times the settlements at the end of the construction phase. Final settlements of 15–35 cm occurred [42,43].

Nearly all high-rise buildings founded on spread foundations in Frankfurt Clay have differential settlements, which lead to a tilting of the superstructures [43]. A statistical evaluation of the measurements indicates that this tilting is up to 20–30% of the average settlement, even if the foundation is loaded centrically [44]. The differential settlements result from the inhomogeneity of the Frankfurt subsoil.

3.4.1 High-rise building complex of Zürich Assurance

The high-rise building complex of the Zürich Assurance Company in Frankfurt am Main, Germany, was constructed from 1959 to 1963. It consists of two towers 63 m and 70 m high, respectively, and an annex building up to eight storeys. The whole complex has two sublevels and is founded on a spread foundation. The foundation depth is 7 m below the surface. The ground view is shown in Figure 3.24.

The soil and groundwater conditions are representative for Frankfurt am Main. At the surface are fillings and quaternary sands and gravel. At a depth of about 7 m below the surface, begins the tertiary Frankfurt Clay, which consists of alternating layers of stiff to semisolid clay and limestone. At a depth of 67 m below the surface follows the Frankfurt Limestone. The groundwater level is about 5–6 m below the surface.

The measured settlements at the end of the construction of the superstructure are about 60% of the final settlements (Figure 3.25). After the end of the construction, the settlement rate decreased due to the consolidation process. About 5 years after the end of the construction, the settlements come to an end at about 8.5–9.5 cm.

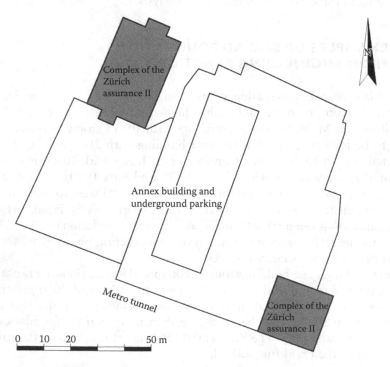

Figure 3.24 Ground view of the high-rise building complex of Zürich Assurance.

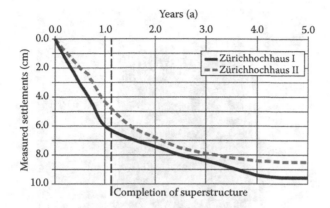

Figure 3.25 Measured settlements.

In the years 2001 and 2002 the high-rise building complex was deconstructed. In its place now is the Opernturm, with a height of 177 m [45,46].

3.4.2 Westend Gate

The high-rise Westend Gate building (former name: Senckenberganlage) in Frankfurt am Main, Germany, was constructed from 1972 to 1974 (Figure 3.26). It is 159 m high and is founded on a spread foundation system. The basement has three sublevels. The building is an office tower up to the 23rd floor. Above the office part is the Marriott Hotel. The soil and groundwater conditions are similar to the high-rise building complex of the Zürich Assurance.

Westend Gate is the high-rise building with the biggest settlements in Frankfurt am Main [47]. The measured settlements of the building were greater than 30 cm, caused by comparatively high contact pressures of 650 kN/m². The raft foundations were only arranged under the high-rise building. The annex sublevels were founded on single foundations (Figure 3.27). For controlling the settlements and the differential settlements, expansion joints were arranged between the foundation elements and the superstructure. The expansion joints were closed after finishing the reinforcedconcrete cores. The flexible steel structure, which reaches from the third to the 23rd floor, was not damaged by the settlements and the differential settlements. The storeys above the 23rd floor were constructed with reinforcedconcrete cells with a comparatively high stiffness. Hydraulic jacks have been installed between the flexible steel construction and the stiff concrete cells. The hydraulic jacks balance the occurring settlements. Due to the long-term settlement behavior of the soil, several joints in the upper floors have been kept open until two years after the construction [47,48].

Figure 3.26 Westend Gate.

3.4.3 Silver Tower

The Silver Tower (formerly Dresdner Bank) in Frankfurt am Main, Germany, is 166 m high and was constructed from 1975 to 1978 (Figure 3.28). The Silver Tower is constructed on a foundation raft with an average thickness of 3.5 m. The foundation level is 14 m deep under the surface. The soil and groundwater conditions are similar to the high-rise building complex of the Zürich Assurance.

Due to the eccentric loading, 22 pressure cushions were installed in the northwest under the foundation raft (Figure 3.29) [42,49]. The pressure cushions have a size of 5 m × 5 m and consist of soft rubber with a thickness of 3 mm. The tightness of the pressure cushions was checked before the installation. The cushions were filled initially with water. The pressures inside the cushions were regulated so only small differential settlements occurred. After the end of the construction and the adjustment of the high-rise building, the water in the cushions was replaced by mortar.

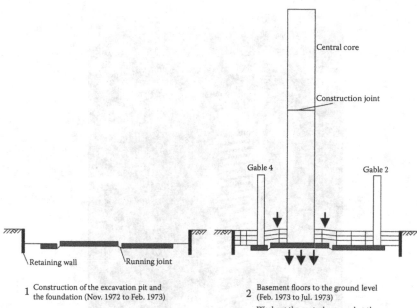

1 Construction of the excavation pit and the foundation (Nov. 1972 to Feb. 1973)

Retaining wall   Running joint

Central core

Construction joint

Gable 4   Gable 2

2 Basement floors to the ground level (Feb. 1973 to Jul. 1973)

Works at the central core and at the gables (Jul. 1973 to Oct. 1973)

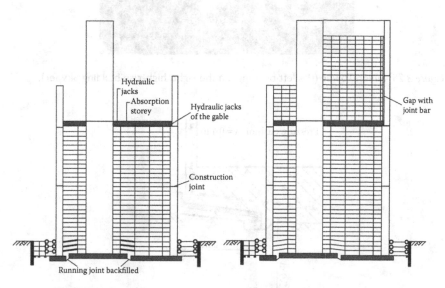

Hydraulic jacks

Absorption storey

Hydraulic jacks of the gable

Construction joint

Running joint backfilled

3 Installation of the floor slabs of the office (Oct. 1973 to Mar. 1974)

Absorption storey and construction of the remaining gables (Mar. 1974 to Jul. 1974)

Gap with joint bar

4 Finishing of construction (Jul. 1974 to Dec. 1974)

Figure 3.27 Construction phases.

Figure 3.28 Silver Tower (the left building; on the right: high-rise building Skyper).

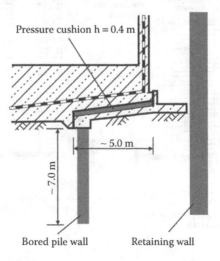

Figure 3.29 Hydraulic devices to adjust the settlements.

3.4.4 Frankfurt Bureau Centre (FBC)

The FBC is a 142 m high-rise building in Frankfurt am Main, Germany, which is founded on 3.5 m thick foundation raft. The foundation level is about 12.5 m below the surface. Figure 3.30 shows the high-rise building from the south. It was constructed from 1973 to 1980. The long construction time was due to a lack of investment during the oil crisis. The soil and groundwater conditions are similar to the high-rise building complex of the Zürich Assurance.

From the beginning of the construction, the settlements were measured for 5 years (Figure 3.31). The maximum final settlement was about 28 cm in the core area of the high-rise building [42]. About 1.5 years after construction ended, the settlements were about 70% of the final settlements. The differential settlements between the high-rise building and the adjacent

Figure 3.30 Frankfurt Bureau Centre (FBC).

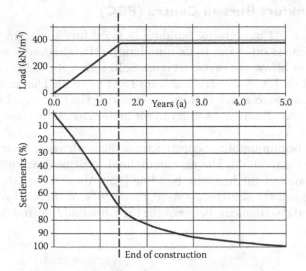

Figure 3.31 Measured settlements.

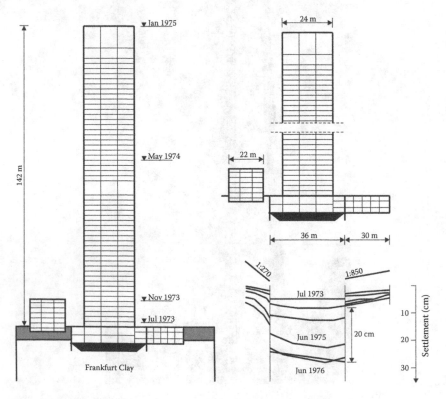

Figure 3.32 Cross section of the structure and measured settlements.

buildings are between 9.5 cm and 20 cm (Figure 3.32). The tilting of the high-rise building is about 1:1350 [50].

3.4.5 Twin towers of Deutsche Bank

The twin towers of the Deutsche Bank in Frankfurt am Main, Germany, are 158 m high and were constructed from 1979 to 1984 (Figure 3.33). The towers are on a foundation raft with a size of 80 m × 60 m and a thickness of 4 m. The foundation level is about 13 m below the surface [51]. The soil and groundwater conditions are similar to the high-rise building complex of the Zürich Assurance.

The measured settlements are between 10 cm and 22 cm. Figure 3.34 shows the isolines of the settlements. To minimize the influence of the twin towers on the adjacent buildings, hydraulic jacks were installed (Figure 3.35). The possible regulation of the differential settlements by the hydraulic jacks was about ± 8 cm.

Figure 3.33 Twin Towers of Deutsche Bank.

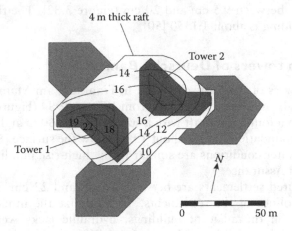

Figure 3.34 Measured settlements.

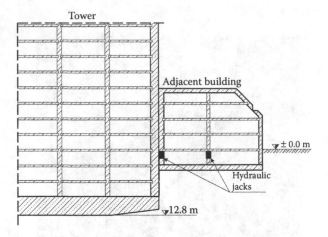

Figure 3.35 Cross-section of the superstructure with hydraulic jacks.

REFERENCES

1. Bundesministerium für Verkehr, Bau und Stadtentwicklung (2012): Richtlinie für die Standardisierung des Oberbaus von Verkehrsflächen (RStO 12).
2. Bundesministerium für Verkehr, Bau und Stadtentwicklung (2009): Zusätzliche Technische Vertragsbedingungen und Richtlinien für Erdarbeiten im Straßenbau (ZTV E-StB 09).
3. Deutsches Institut für Normung e.V. (2001): DIN EN ISO 13793 Thermal Performance of Buildings: Thermal Design of Foundations to Avoid Frost Heave. Beuth Verlag, Berlin.
4. Deutscher Ausschuss für Stahlbeton e.V. (2003): DAfStb-Richtlinie Wasserundurchlässige Bauwerke aus Beton (WU-Richtlinie). Beuth Verlag, Berlin.

5. Deutscher Ausschuss für Stahlbeton e.V. (2006): Heft 555 Erläuterungen zur DAfStb-Richtlinie Wasserundurchlässige Bauwerke aus Beton. Beuth Verlag, Berlin.
6. Lohmeyer, G.; Ebeling, K. (2013): Weiße Wannen einfach und sicher. 10. Auflage, Verlag Bau + Technik, Düsseldorf, Germany.
7. Haack, A.; Emig, K.-F. (2003): Abdichtungen im Gründungsbereich und auf genutzten Deckenflächen. 2. Auflage, Ernst & Sohn Verlag, Berlin.
8. Deutsches Institut für Normung e.V. (2014): DIN EN 1997-1 Eurocode 7: Geotechnical design: Part 1: General rules. Beuth Verlag, Berlin.
9. Deutsches Institut für Normung e.V. (2010): DIN EN 1997-1/NA National Annex: Nationally Determined Parameters—Eurocode 7: Geotechnical Design–Part 1: General Rules. Beuth Verlag, Berlin.
10. Deutsches Institut für Normung e.V. (2010): DIN 1054 Subsoil: Verification of the Safety of Earthworks and Foundations—Supplementary Rules to DIN EN 1997-1. Beuth Verlag, Berlin.
11. Deutsches Institut für Normung e.V. (2012): DIN 1054 Subsoil: Verification of the Safety of Earthworks and Foundations—Supplementary rules to DIN EN 1997-1:2010; Amendment A1:2012. Beuth Verlag, Berlin.
12. Deutsches Institut für Normung e.V. (2009): DIN 4084 Soil: Calculation of Embankment Failure and Overall Stability of Retaining Structures. Beuth Verlag, Berlin.
13. Deutsches Institut für Normung e.V. (2012): DIN 4084 Ground: Calculation of the Overall Stability—Supplement 1: Examples of calculation. Beuth Verlag, Berlin.
14. Hettler, A. (2000): *Gründung von Hochbauten*. Ernst & Sohn Verlag, Berlin.
15. Deutsches Institut für Normung e.V. (1974): DIN 4018 Subsoil: Contact Pressure Distribution Under Raft Foundations, Analysis. Beuth Verlag, Berlin.
16. Deutsches Institut für Normung e.V. (1981): DIN 4018 Supplement 1 Subsoil: Analysis of Contact Pressure Distribution Under Raft Foundations; Explanations and Examples of Analysis. Beuth Verlag, Berlin.
17. Boussinesq, M.J. (1885): Application des Potentials à l'Étude de l'Équilibre et du Mouvement des Solides Élastiques. Gauthier-Villard, Paris, France.
18. Katzenbach, R.; Zilch, K.; Moormann, C. (2012): *Baugrund-Tragwerk-Interaktion. Handbuch für Bauingenieure: Technik, Organisation und Wirtschaftlichkeit*. Springer Verlag, Heidelberg, Germany, 1471–1490.
19. Kany, M. (1959): *Berechnung von Flächengründungen*. Ernst & Sohn Verlag, Berlin.
20. Kany, M. (1974): *Berechnung von Flächengründungen, Band 2, 2. Auflage*, Ernst & Sohn Verlag, Berlin.
21. Meyerhof, G.G. (1979): General report: Soil–structure interaction and foundations. 6th Panamerican Conference on Soil Mechanics and Foundation Engineering, 2–7 December, Lima, Peru, 109–140.
22. Borowicka, H. (1943): Über ausmittig belastete starre Platten auf elastisch-isotropem Untergrund. *Ingenieur-Archiv*, XIV. Band, Heft 1, Springer Verlag, Berlin, 1–8.
23. Lang, H.J.; Huder, J.; Amann, P. (2003): *Bodenmechanik und Grundbau. 7. Auflage*, Springer Verlag, Berlin.

24. Smoltczyk, U.; Vogt, N. (2009): Flachgründungen. Grundbautaschenbuch, Teil 3: *Gründungen und geotechnische Bauwerke*. 7. Auflage, Ernst & Sohn Verlag, Berlin, 1–71.
25. Winkler, E. (1867): *Die Lehre von der Elastizität und Festigkeit*. Verlag Dominicus, Prague, Czech Republic.
26. Ohde, J. (1942): Die Berechnung der Sohldruckverteilung unter Gründungskörpern. *Der Bauingenieur* 23, Germany, Heft 14/16, 99–107 and 122–127.
27. Deutsches Institut für Normung e.V. (2005): DIN 1054 Subsoil: Verification of the Safety of Earthworks and Foundations. Beuth Verlag, Berlin.
28. Katzenbach, R.; Boled-Mekasha, G.; Wachter, S. (2006): Gründung turmartiger Bauwerke. *Beton-Kalender*, Ernst & Sohn Verlag, Berlin, 409–468.
29. Deutsches Institut für Normung e.V. (2006): DIN 4017 Soil: Calculation of Design Bearing Capacity of Soil Beneath Shallow Foundations. Beuth Verlag, Berlin.
30. Deutsches Institut für Normung e.V. (2006): DIN 4017 Soil: Calculation of Design Bearing Capacity of Soil Beneath Shallow Foundations—Calculation Examples. Beuth Verlag, Berlin.
31. Prandtl, L. (1920): Über die Härte plastischer Körper. Nachrichten von der Königlichen Gesellschaft der Wissenschaften zu Göttingen. *Mathematische Klasse*, Berlin.
32. Deutsches Institut für Normung e.V. (2011): DIN 4019 Soil: Analysis of Settlement. Beuth Verlag, Berlin.
33. Arbeitskreis Berechnungsverfahrender Deutschen Gesellschaft für Erd- und Grundbau e.V. (1993): Empfehlungen Verformungen des Baugrund bei baulichen Anlagen: EVB. Ernst & Sohn Verlag, Berlin.
34. Skempton, A.W.; MacDonald, D.H. (1956): The allowable settlements of buildings. *Proceedings of the Institute of Civil Engineering*, 10 May, London, Great Britain, 727–783.
35. Bjerrum, L. (1973): Allowable settlements of structures. Norwegian Geotechnical Institute, Publication Nr. 98, Oslo, Norway, 1–3.
36. Schultze, E.; Muhs, H. (1967): Bodenuntersuchungen für Ingenieurbauten. 2. Auflage, Springer Verlag, Berlin.
37. Ziegler, M. (2012): Geotechnische Nachweise nach EC 7 und DIN 1054: Einführung mit Beispielen. 3. Auflage, Wilhelm Ernst & Sohn, Berlin.
38. Dörken, W.; Dehne, E.; Kliesch, K. (2012): Grundbau in Beispielen Teil 2. 5. Auflage, Werner Verlag, Neuwied, Germany.
39. Deutsches Institut für Normung e.V. (2011): DIN 18196 Earthworks and Foundations: Soil Classification for Civil Engineering Purposes. Beuth Verlag, Berlin.
40. Deutsches Institut für Normung e.V. (1996): DIN 18126 Soil, investigation and Testing: Determination of Density of Non-cohesive Soils for Maximum and Minimum Compactness. Beuth Verlag, Berlin.
41. Deutsches Institut für Normung e.V. (2012): DIN 18127 Soil, investigation and Testing: Proctor-test. Beuth Verlag, Berlin.
42. Sommer, H (1976): Setzungen von Hochhäusern und benachbarten Anbauten nach Theorie und Messungen. *Vorträge der Baugrundtagung* in Nürnberg, Germany, 141–169.

43. Sommer, H. (1978): Messungen, Berechnungen und Konstruktives bei der Gründung Frankfurter Hochhäuser. *Vorträge der Baugrundtagung* in Düsseldorf, Germany, 205–211.

44. Sommer, H.; Tamaro, G.; DeBeneditis, C. (1991): Messeturm, foundations for the tallest building in Europe. *Proceedings of 4th International Conference on Piling and Deep Foundations*, April, Stresa, Italy, 139–145.

45. Katzenbach, R.; Leppla, S.; Seip, M. (2011): Das Verformungsverhalten des Frankfurter Tons infolge Baugrundentlastung. *Bauingenieur* 86, May, Springer VDI Verlag, Düsseldorf, Germany, 233–240.

46. Katzenbach, R.; Leppla, S. (2013): Deformation behaviour of clay due to unloading and the consequences on construction projects in inner cities. *18th Conference of the International Society for Soil Mechanics and Geotechnical Engineering*, 2–6 September, Paris, France, Vol. 3, 2023–2026.

47. Katzenbach, R. (1995): Hochhausgründungen im setzungsaktiven Frankfurter Ton. 10. Christian Veder Kolloquium, 20 April, Graz, Austria, 44–58.

48. Moos, G. (1976): Hochhaus Senckenberganlage in Frankfurt am Main. Ph. Holzmann AG, Technischer Bericht, Frankfurt, Germany, 1–25.

49. Gravert, F.W. (1975): Ein Beitrag zur Gründung von Hochhäusern auf bindigen Böden. Deutsche Konferenz Hochhäuser, Deutsche Gruppe der Internationalen Vereinigung für Brückenbau und Hochbau, 2–4 October, Mainz, Germany, 216–224.

50. Stroh, D.; Katzenbach, R. (1978): Der Einfluss von Hochhäusern und Baugruben auf die Nachbarbebauung. Bauingenieur 53, Springer-Verlag, Berlin, 281–286.

51. Katzenbach, R.; Bachmann, G.; Boled-Mekasha, G.; Ramm, H. (2005): Combined Pile-Raft Foundations (CPRF): An appropriate solution for the foundation of high-rise buildings. *Slovak Journal of Civil Engineering*, No. 3, 19–29.

Chapter 4

Deep foundations

For the transfer of high loads into the subsoil as well as for spread foundations on subsoil with insufficient stiffness, deep foundations are designed. These deep foundations include the following types:

- Pile foundations
- Barrette foundations
- Combined pile-raft foundations (CPRF) *
- Well foundations *
- Caisson foundations *

This chapter only deals with the classic deep foundation constructed out of piles and barrettes. The deep foundation systems marked with an asterisk (*) are special forms that are discussed in other chapters. Barrettes are single diaphragm wall lamellas and can be used analog as deep foundation elements [1].

For deep foundations, a multitude of technical regulations exist. Thus, in this chapter only the most significant regulations for design, construction and quality assurance can be mentioned. These regulations only contain basic information and definitions. Furthermore, new developments mostly should not be adopted into the standards so that a large number of complementing essays exist. In the present case of deep foundations the study-group "Piles" of the German Society for Geotechnics e.V. (DGGT) should be emphasized particularly. Analogous to other study-groups of the DGGT, the study-group "Piles" has compiled an addition to the current standards in [2], which has to be counted among the technical regulations. Fundamental for the planning, design and construction are the regulations [1,3–8].

4.1 PILE TYPES

In practice, different pile types are applied.

- Bored piles
- Driven piles

- Wood piles
- Steel piles
- Driven reinforced-concrete piles
- Driven *in situ* concrete piles
- Driven bored piles (screw piles)
- Micro piles

Table 4.1 shows the advantages and disadvantages of the different pile types in accordance with [2,9]. The choice of a pile type depends on the following criteria:

- Structural loads
- Location, geometry, and neighborhood
- Subsoil and groundwater conditions
- Deformation limits of the construction
- Economics
- Availability of construction materials
- Availability of construction machines
- Availability of a specialized heavy construction company

The single pile types are characterized in [2]. A general distinction is made between bored and driven piles. Driven bored piles, also named as screw piles, are a combination of both and can be divided into partial- and full-driven bored piles.

A bored pile is constructed by soil excavation. The generated cavity is refilled with concrete and a reinforcement. Driven piles include piles made out of wood, steel, reinforced concrete, and driven *in situ* concrete. During the installation of driven piles, the surrounding subsoil is displaced and compressed. During the installation of partial-driven bored piles, parts of the subsoil are excavated and the residual subsoil is displaced laterally. During the installation of full-driven bored piles, the surrounding subsoil is displaced completely. Comparable to classic driven piles, an increase of the density and thus a favorable bearing capacity and a favorable load deformation behavior can be observed in fully driven bored piles. In [10] the load deformation behavior of *in situ* concrete piles with variable soil displacement is investigated.

4.2 CONSTRUCTION

The construction of deep foundations is one of the classic tasks of special heavy construction projects. Figure 4.1 shows the construction of bored piles in the inner city of Frankfurt am Main, Germany. Special construction machines and experienced personnel are required for a successful construction. The supervision of the construction works on site by specially qualified persons is necessary in most instances.

Table 4.1 Advantages and disadvantages of different pile types

Advantages	Disadvantages
Bored piles	
• Vibrationless pile production • Soil exploration owing to boring • No restrictions because of working height or neighboring during the construction process • Great depth and diameter is possible	• Pile inclination limited to ca. 10:1 • Extraction of the piping involves the risk of damage of the green concrete coumn and the armor • Soil material can be irrupted as a result of quickly pulling the piping
Wood piles	
• High degree of elasticity • Easy in processing • Long life performance under water • Relatively economical	• Quick destruction through putridity owing to air admission • Limited driveability in heavy soils • Low load-bearing capacity and length compared to other pile types
Steelpiles	
• High strength and elasticity • Big range of profiles • Insensitivity during transport • Extension easily possible • Good connection options • Inclinations up to 1:1	• Relatively high cost of material • Risk of corrosion • Risk of aeolian corrosion • Risk of loss of the position stability, respectively, risk of rotation of the profile during the pile driving • Noise and vibration during the pile driving/vibration
Reinforced-concrete driven piles	
• Resistant to sea water • Good soil compaction during the pile driving • Good connection options between the pile and the construction • Inclinations up to 1:1 • Relatively economical	• Heavy and clumsy • Sensitively to bending, e.g., during transport • Risk of cracks in consequence of incorporation and installation • Heavy pile drivers required • Noise and vibration during the pile driving
In situ **concrete displacement piles**	
• Good compaction of the surrounding subsoil from which a high load-carrying capacity results	• Noise and vibration during the pile driving • Risk of damage to the fresh neighboring piles • Limited inclination • Lost foot plate
Full displacement bored piles (screw piles)	
• Good compaction of the surrounding subsoil from which a high load-carrying capacity results • Vibrationless pile production	• Inclination up to 4:1 • Lost tip
Micro piles	
• Also producible under highly constrained spaces	• Risk of buckling because of very small diameter

Figure 4.1 Construction of a bored pile in the inner city of Frankfurt am Main, Germany.

The technical development of pile types and pile dimensions in recent years seems to be unlimited. The present pile types and the necessary construction machines are being continuously improved, and new systems and machines are being developed. For example, bored piles with a diameter of 3 m and a length of more than 80 m can now be constructed [11]. A good overview of the technological equipment as well as of the different pile types is given in [12–14].

The essential technical standards for advertisement and construction, as well as for quality control, for the construction of deep foundations are

- DIN EN 1536 [15] and DIN SPEC 18140 [16] for bored piles
- DIN EN 12794 [17] for foundation piles constructed out of precast concrete elements
- DIN EN 12699 [18,19] and DIN SPEC 18538 [20] for driven piles
- DIN EN 14199 [21] and DIN SPEC 18539 [22] for micro piles
- DIN EN 12063 [23] for deep foundations constructed out of barrettes
- DIN 18301 [24] for drilling operations
- DIN 18304 [25] for driving, vibrating, and pressing work
- DIN 4126 [26], DIN 4126 supplement 1 [27], DIN 4127 [28], DIN EN 1538 [29], DIN 18313 [30] for diaphragm wall work

The construction of bored piles should always be operated by applying a casing and/or a stabilization fluid, for example, water, bentonite, or polymer slurries. The casing is pressed into the subsoil using a hydraulic casing machine with oscillating rotational motion. The subsoil is excavated by a grabber, drilling bucket, or similar techniques. The casing has to be ahead of the excavation down to the pile base. A sufficient surcharge against the surrounding groundwater must be ensured at any time of the construction of bored piles. The drilling has to be constructed with a sufficient surcharge to prevent an entry of soil as a result of hydraulic failure and to prevent a bulking around the pile. According to [15], the water level in the borehole should be 1.5 m above the level of the surrounding groundwater. Figure 4.2 shows the construction process of a bored pile. During the pulling out of the casing, the concrete level has to be higher than the edge of the casing. In the majority of cases, the pile head has an insufficient strength in the first 50 cm. The concrete of this weakness zone has to be removed and has to be concreted anew, together with the connecting elements. The construction of bored piles has to be supervised on site by a geotechnical specialist. Continuously, the water level in the borehole, the borehole depth, the depth of the casing, and the concrete level have to be documented.

If piles are constructed by driving or vibrating the observance of the neighborhood is essential because of the noise and agitation emissions. If necessary, extensive measurements according to the observational method have to be carried out.

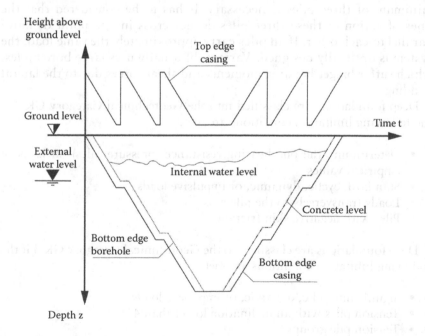

Figure 4.2 Construction process of a bored pile.

At the construction of micro piles, it must be guaranteed that the grouting works do not induce soil displacements in unintended areas. Therefore, the grouting work must be supervised strictly and monitored precisely.

4.3 GEOTECHNICAL ANALYSIS

4.3.1 Basics

Deep foundation elements transfer the loads into deeper soil layers that normally have a higher bearing capacity and a bigger stiffness. Deep foundations are a comparatively robust foundation type with small displacements. Load transfer into the soil or rock occurs via the skin friction at the pile shaft and/or via the base resistance under the pile toe. The load transfer via a raft or a strip foundation grid at the pile top is not taken into account.

For horizontal loads two types of deep foundation are possible (Figure 4.3). At variant (a) the transfer of horizontal loads is realized by inclined piles. At variant (b) the transfer of horizontal loads is realized by a horizontal bedding. Normally, the connection between the piles and the raft or strip foundation is seen as a flexible joint. To achieve a sufficient stability of the construction in a two-dimensional view, a minimum of three piles is necessary. It has to be considered that the lines of action of these three piles do not cross in one point or move parallel to each other. If all piles carry approximately the same load, the system is optimally designed. Variant (b) usually uses large bored piles, which suffer bigger bending moments and shear forces due to the lateral bedding.

Deep foundations are classified into the Geotechnical Category GK 2 if the following limitation conditions are met.

- Determining the pile bearing resistance (pressure piles) on basis of empirical values
- Standard, cyclic, dynamic, or impulsive loads
- Loads transversely to the pile axis
- Piles with negative skin friction

Deep foundations are classified to the Geotechnical Category GK 3 if the following limitation conditions are met.

- Significant cyclic, dynamic, or impulsive loads
- Tension piles with an inclination lower than 45°
- Tension pile groups
- Grouted pile systems

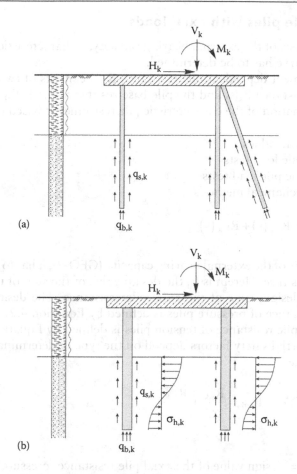

Figure 4.3 Load transfer at pile foundations. (a) Inclined piles; (b) large drilled piles.

- Determining the pile bearing resistance (tension piles) on basis of empirical values
- Loads transversely to the pile axes from lateral pressure or settlements
- Piles with very high loads in combination with strongly restricted settlements
- Piles with shaft and/or pile toe grouting

The analysis of the external bearing capacity is done according the limit state GEO-2. The analysis of the internal bearing capacity is done according to the limit state STR. However, the analysis of the internal bearing capacity depends on the material and is defined by the associated standards and regulations. In the following, the focus is on the geotechnical analysis (external bearing capacity).

4.3.2 Single piles with axial loads

For the analysis of the external bearing capacity, a characteristic resistance settlement curve has to be determined.

The settlement dependent pile resistance $R_c(s)$ consists of two parts: the pile shaft resistance $R_s(s)$ and the pile base resistance $R_b(s)$ (Equation 4.1). The determination of the characteristic pile resistance is based on

- Empirical values
- Static pile load tests
- Dynamic pile load tests
- Soil mechanical methods

$$R_{c,k}(s) = R_{s,k}(s) + R_{b,k}(s) \tag{4.1}$$

For analysis of the external bearing capacity (GEO-2), it has to be checked that $F_d \leq R_d$ is true. Herein is F_d the design value of the sum of the impacts and R_d the design value of the axial pile resistance. The design value of the pile resistance of pressure piles is defined by Equation 4.2. The design value of the pile resistance of tension piles is defined by Equation 4.3. The necessary partial safety factors depend on the type of determination of the pile resistance.

$$R_{c,d} = \frac{R_{c,k}}{\gamma_t} \text{ or } R_{c,d} = \frac{R_{s,k}}{\gamma_s} + \frac{R_{b,k}}{\gamma_b} \tag{4.2}$$

where:

$R_{c,d}$ = design value of the axial pile resistance (pressure)
$R_{c,k}$ = characteristic axial pile resistance (pressure)
$R_{s,k}$ = pile skin resistance
$R_{b,k}$ = pile base resistance
$\gamma_s, \gamma_b, \gamma_t$ = partial safety factor in accordance to the type of determination of the pile resistance

$$R_{t,d} = \frac{R_{t,k}}{y_{s,t}} \tag{4.3}$$

where:

$R_{t,d}$ = design value of the axial pile resistance (tension)
R_t = R_s = characteristic axial pile resistance (tension)
$\gamma_{s,t}$ = partial safety factor of the pile skin resistance

If one of the two reaction effects prevails, the piles are indicated as base pressure piles or skin friction piles (Figure 4.4) [31].

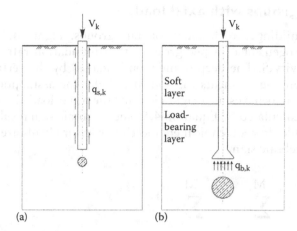

Figure 4.4 Friction pile and base pressure pile. (a) Friction pile; (b) base pressure pile. (Pulsfort, M. Grundbau, Baugruben und Gründungen. Handbuch für Bauingenieure: Technik, Organisation und Wirtschaftlichkeit, Springer-Verlag, Heidelberg, Germany, 2012, 1568–1639.)

The self-weight can be disregarded at piles with axial pressure load. The self-weight can be included in the analysis of piles with axial tensile load.

For abandoned piles in soft soils with axial pressure load, the safety against buckling must be verified if the undrained cohesion is $c_u \leq 15$ kN/m^2.

Two different failure mechanisms must be investigated for tension piles. On the one hand, the pull-out resistance of each pile must be checked. On the other hand, the safety against uplift of the entire subsoil block, which surrounds the tension piles, has to be checked. The second analysis is reflected in Chapter 4.3.3. The stress of tension piles is defined in Equation 4.4:

For determining pile shaft resistance, a pile load test is applied [2]:

$$F_{t,d} = F_{t,G,k} \cdot \gamma_G + F_{t,Q,rep} \cdot \gamma_Q - F_{c,G,k} \cdot \gamma_{G,inf} \qquad (4.4)$$

where:

$F_{t,G,k}$ = characteristic value of the tensile load as a result of dead loads

γ_G = partial safety factor for dead loads in the limit state GEO-2

$F_{t,Q,rep}$ = characteristic resp. representative value of the tensile load as a result of unfavorable variable effects

γ_Q = partial safety factor for unfavorable loads in the limit state GEO-2

$F_{c,G,k}$ = characteristic value of a simultaneously acting pressure load as a result of constant loads

$\gamma_{G,inf}$ = partial safety factor for constant pressure load in the limit state GEO-2

4.3.3 Pile groups with axial loads

Normally, buildings are founded on pile groups. Figure 4.5 shows an optional arrangement of a pile group with a foundation raft around the center of gravity S. The deep foundation is loaded by the vertical impact V and the bending moments M_y and M_x. Under the assumption that the foundation raft behaves as a rigid element, the axial load $E_{i,k}$ on a single pile i can be calculated by Equation 4.5. The algebraic sign results from the load on the pile. In geotechnical engineering, pressure loads are defined by a positive algebraic sign.

$$E_{i,k} = \pm \frac{V}{n} \pm \frac{M_y}{\sum x_i^2} \cdot x_i \pm \frac{M_x}{\sum y_i^2} \cdot y_i \qquad (4.5)$$

The load deformation behavior of a pile group cannot be estimated based on a load-settlement curve of a single pile because the single piles of a pile group influence each other [9]. For an approximate analysis of the settlements, the pile group can be considered as a deep spread foundation. For homogenizing the settlements of a pile group, the connecting foundation raft, or the strip foundation grid must be sufficiently rigid.

Tension pile groups are often used for buoyancy control of excavations, docks, or locks. The piles form a compact subsoil block with the surrounding soil (Figure 4.6). For analysis of the safety against uplift, the buoyancy of the subsoil block (limit state UPL) and the Equations 4.6 and 4.7 are used. The weight $G_{E,k}$ of the trailed subsoil block of a single tension pile has to be determined by the weight γ of the soil (Figure 4.7). For soils under the groundwater level, the weight γ' must be applied.

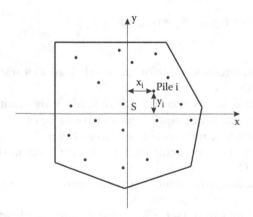

Figure 4.5 Optional arrangement of a pile group.

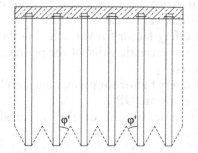

Figure 4.6 Subsoil block of a tension pile group.

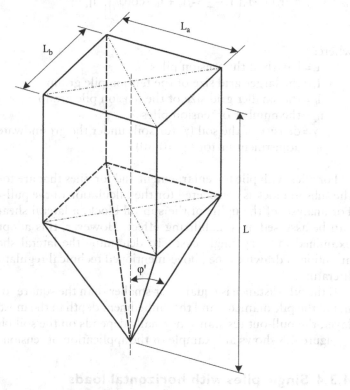

Figure 4.7 Geometry of a subsoil block of a single pile.

$$G_{dst,k} \cdot \gamma_{G,dst} + Q_{dst,rep} \cdot \gamma_{Q,dst} \leq G_{stb,k} \cdot \gamma_{G,stb} + G_{E,k} \cdot \gamma_{G,stb} \qquad (4.6)$$

where:

$G_{dst,k}$ = characteristic value of the permanent destabilizing vertical effects

$\gamma_{G,dst}$ = partial safety factor for permanent stabilizing effects in the limit state UPL

$Q_{dst,rep}$ = characteristic resp. representative value of the variable destabilizing vertical loads
$\gamma_{Q,dst}$ = partial safety factor for variable destabilizing vertical loads
$G_{stb,k}$ = lower characteristic value of the permanent stabilizing vertical effects
$\gamma_{G,stb}$ = partial safety factor for permanent stabilizing vertical effects
in the limit state UPL
$G_{E,k}$ = characteristic weight of the attached soil (tension pile
group)

$$G_{E,k} = n_z \left[l_a \cdot l_b \left(L - \frac{1}{3} \cdot \sqrt{l_a^2 + l_b^2} \cdot \cot\varphi' \right) \right] \cdot \eta_z \cdot \gamma \qquad (4.7)$$

where:
L = length of the tension piles
l_a = the larger grid size of the tension pile group
l_b = the smaller grid size of the tension pile group
n_z = the number of tension piles
γ = density of the soil (γ' for soils under the groundwater table)
η_z = adjustment factor ($\eta_z = 0.80$)

For piles with pile toe enlargement and for piles that are founded in rock, the subsoil block is considered for the calculation of the pull-out resistance. For analysis of the uplift of the subsoil block, a lateral shear resistance T_d can be assessed as a stabilizing effect. However, this adoption has to be examined in every single case. To determine the lateral shear resistance, attention is drawn to the above-mentioned technical regulations and to the literature.

If the pile distance is equal to or smaller than the square root of the product of the pile diameter and the embedment depth in the mostly bearing soil layer, the pull-out resistance normally depends on the soil block.

Figure 4.8 shows an example of the application of tension pile groups.

4.3.4 Single piles with horizontal loads

According to [1], piles with a diameter $D_s \geq 0.30$ m and an edge length $a_s \geq 0.30$ m may be used for carrying horizontal loads in a foundation system. The horizontal resistance of a single pile can be characterized by the horizontal subgrade reaction modulus $k_{s,k}$. The soil stresses cannot become larger than the passive earth pressure e_{ph}.

For the determination of the internal forces, Equation 4.8 can be used to calculate the horizontal subgrade reaction modulus. Therefore, the stiffness modulus $E_{s,k}$ has to be determined by laboratory or field tests if no pile load tests *in situ* are performed. If the following equation is used, the

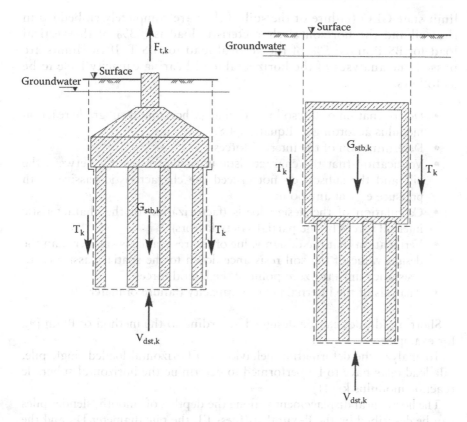

Figure 4.8 Application of tension piles for the protection against buoyancy.

horizontal displacement at the pile head has to be limited to 0.03 D_s to maximal 2 cm:

$$k_{s,k} = \frac{E_{s,k}}{D_s} \qquad (4.8)$$

where:

$k_{s,k}$ = horizontal subgrade reaction modulus for a stress level under characteristic resp. representative loads

$E_{s,k}$ = stiffness modulus for a stress level under characteristic resp. representative loads

D_s = pile diameter (resp. pile edge length a_s)

To ensure the stability of horizontal-loaded piles, $F_{tr,d} \le R_{tr,d}$ has to be verified. Smooth, slender piles must be analyzed according to the limit state STR (failure of the structure or its elements) and according to the limit state GEO (failure of the soil) if they are completely embedded in the soil and the horizontal characteristic load is $\leq 3\%$ of the vertical load for BS-P and $\leq 5\%$ of the vertical load for BS-T. If the limits are passed, the analyses of the horizontal load-bearing capacity have to be as follows:

- Determination of the soil reaction (e.g., horizontal subgrade reaction modulus according to Equation 4.8)
- Determination of the internal forces
- Verification that the characteristic normal stresses $\sigma_{h,k}$ between the pile and the subsoil do not exceed the characteristic passive earth pressure $e_{ph,k}$ at any point
- Calculation of the design loads (factorization of the characteristic internal forces by the partial safety factors)
- Verification that the design value of the resistance is smaller than the design value of the soil resistance down to the spatial passive earth resistance until the zero point of the lateral forces
- Analysis of the internal bearing capacity (failure of material)

Short, rigid piles can be designed according to the method of Blum [9], for example.

To analyze the deformation behavior of a horizontal-loaded single pile, pile load tests have to be performed to determine the horizontal subgrade reaction modulus $k_{s,k}$ [1].

The horizontal displacement y along the depth z of smooth, slender piles can be described by the flexural stiffness EI, the pile diameter D_s, and the differential equation of an elastic beam considering the horizontal stress σ_h (Equation 4.9). Analogous to the subgrade reaction modulus method, the Winkler spring model specifies the relation between horizontal pile displacement y and the horizontal stress σ_h, which can be determined approximately by $\sigma_h = k_s\, y$ [9].

$$k_{s,k,i}\left(z\right) = k_{s,k,i} \cdot \frac{z}{D_s} \tag{4.9}$$

4.3.5 Pile groups with horizontal loads

If pile groups under a raft or under a pile grid are loaded horizontally, all single piles have nearly the same displacement at the pile top. Due to the distinct interaction between the piles, the construction and the subsoil (soil–structure interaction) the single piles get different horizontal loads, depending on the position inside the pile group. For example, the front pile row gets larger horizontal loads than the back pile row [9].

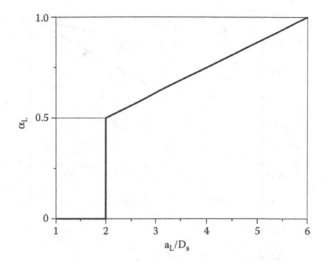

Figure 4.9 Reduction factor α_L in accordance with the relationship between the pile axis in the direction of the force and the diameter of the pile shaft (a_L/D_s).

The distribution of a horizontal load H_G of a double-symmetric pile group can be determined by Equation 4.10. The reduction factors α_i depend on the pile spacing a_L in longitudinal direction, on the pile spacing a_Q in cross direction, as well as on the position of the pile inside the pile group. The reduction factors α_L, α_{QA} and α_{QZ} are defined in the Figures 4.9 and 4.10. The reduction factor α_i is defined in Figure 4.11. At pile spacings of $a_L \geq 6D_s$ and $a_Q \geq 3D_s$, no group effects exist.

$$\frac{H_i}{H_G} = \frac{\alpha_i}{\sum \alpha_i} \qquad (4.10)$$

where:
H_i = horizontal load on pile i
H_G = amount of the horizontal loads on the pile group

For the design of the piles and for calculation of the deformations, the horizontal subgrade reaction modulus must be reduced. The reduction of the horizontal subgrade reaction modulus for bored piles in consolidated cohesive and non-cohesive subsoil as well as in over-consolidated cohesive subsoil is described in [2].

For bored piles in consolidated cohesive and non-cohesive subsoil, the horizontal subgrade reaction modulus can be adopted approximately as

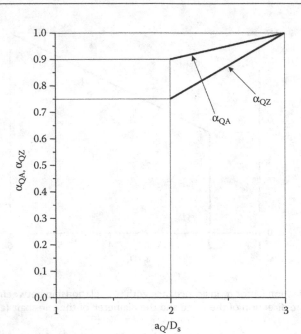

Figure 4.10 Reduction factor α_{QA} and α_{QZ} in accordance with the relationship between the distance of the pile axis transverse to the direction of the force and the diameter of the pile shaft (a_Q/D_s).

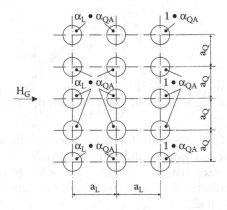

Figure 4.11 Reduction factor α_i in accordance with the location of the pile within the pile group.

linear, increasing with the depth z (Equations 4.10 and 4.11). In Equations 4.10 and 4.11, the following variables are used:

$$L_E = \left(\frac{EI}{k_{sE,k}} \right)^{0.2}$$

$$\text{for} \quad \frac{L}{L_E} \geq 4: \quad k_{s,k,i} = \alpha_i^{1.67} \cdot k_{sE,k}$$

$$\text{for} \quad \frac{L}{L_E} \leq 2: \quad k_{s,k,i} = \alpha_i \cdot k_{sE,k} \tag{4.11}$$

EI = flexural stiffness of the pile i

$k_{sE,k}$ = horizontal subgrade reaction modulus of a single pile in the depth $z = D_s$

$k_{s,i,k}$ = horizontal subgrade reaction modulus of a pile i of the pile group in the depth $z = D_s$

L_E = elastic length of a single pile

L = length of a single pile

The values in the range $4 > L/L_E > 2$ can be interpolated linearly.

For bored piles in over-consolidated cohesive subsoil, the horizontal subgrade reaction modulus can be adopted approximately as constant with the depth (Equations 4.12 and 4.13). The values in the range $4 > L/L_E > 2$ can be interpolated linearly.

$$k_{s,k,i}(z) = k_{s,k,i} = \text{const.} \tag{4.12}$$

$$L_E = \left(\frac{EI}{k_{sE,k} \cdot D_s} \right)^{0.25}$$

$$\text{for} \quad \frac{L}{L_E} \geq 4: \quad k_{s,k,i} = \alpha_i^{1.33} \cdot k_{sE,k}$$

$$\text{for} \quad \frac{L}{L_E} \leq 2: \quad k_{s,k,i} = \alpha_i \cdot k_{sE,k} \tag{4.13}$$

4.3.6 Empirical values for axial loaded piles

The characteristic pile resistance can be conducted by general empirical values if the piles are sufficiently deep integrated into the subsoil. The corresponding regulations for different pile types are contained in [1,3].

In general, the mobilized pile base resistance and the mobilized pile shaft resistance are related to the vertical displacements, which are described by the settlement s at the pile head. Figure 4.12 shows a characteristic resistance settlement curve $R_k(s)$ as well as its parts, which are the base resistance $R_{b,k}(s)$ and the skin friction $R_{s,k}(s)$ until the settlement limit s_g. The settlement limit s_g for the pile base resistance $R_{b,k}(s_g)$ is defined by $s_g = 0.10 \cdot D_s$. The settlement limit s_{sg} for the pile shaft resistance $R_{s,k}(s_{sg})$ is defined in Equation 4.14, with the settlement limit s_{sg} in centimeters and the pile shaft resistance $R_{s,k}(s_{sg})$ in MN.

$$s_{sg} = 0.50 \cdot R_{s,k}\left(s_{sg}\right) + 0.50 \leq 3.0 \, \text{cm} \tag{4.14}$$

The characteristic axial pile resistance $R_k(s)$ can be determined by

$$R_k\left(s\right) = R_{b,k}\left(s\right) + R_{s,k}\left(s\right) = q_{b,k} \cdot A_b + \sum_{i=1}^{i=n} q_{s,k,i} \cdot A_{s,i} \tag{4.15}$$

where:
$R_{b,k}(s)$ = pile base resistance
$q_{b,k}(s)$ = base resistance
A_b = pile base area
$R_{s,k}(s)$ = pile shaft resistance
$q_{s,k,i}$ = skin friction in soil layer i
$A_{s,i}$ = ski area in soil layer i

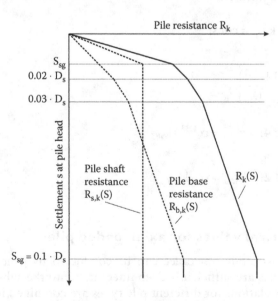

Figure 4.12 Resistance settlement curve.

Tables 4.2 through 4.5 contain the characteristic values of the base resistance $q_{b,k}$ and the values of the skin friction $q_{s,k}$ in non-cohesive and cohesive subsoil. Table 4.6 contains the characteristic values of the base resistance and the skin friction in rock. The table values depend on

- Undrained shear strength c_u of cohesive subsoil
- Cone resistance q_c of a cone penetration test in non-cohesive subsoil
- Unconfined compression strength q_u in rock

Table 4.2 Base resistance $q_{b,k}$ for non-cohesive soils

Relative settlement of the pile head s/D_s resp. s/D_b	Base resistance $q_{b,k}$ [kN/m²] Cone resistance q_c [MN/m²]		
	7.5	15	25
0.02	550–800	1,050–1,400	1,750–2,300
0.03	700–1,050	1,350–1,800	2,250–2,950
0.10($\hat{=} S_g$)	1,600–2,300	3,000–4,000	4,000–5,300

Note: Intermediate values can be interpolated linearly.

For bored piles with an enlarged base, the values must be reduced to 75%.

Table 4.3 Skin friction $q_{s,k}$ for non-cohesive soils

Cone resistance q_c [MN/m²]	Skin friction $q_{s,k}$ [kN/m²]
7.5	55–80
15	105–140
≥ 25	130–170

Note: Intermediate values can be interpolated linearly.

Table 4.4 Base resistance $q_{b,k}$ for cohesive soils

Relative settlement of the pile head s/D_s resp. s/D_b	Base resistance $q_{b,k}$ [kN/m²]		
	Shear strength $c_{u,k}$ of the undrained soil [kN/m²]		
	100	150	250
0.02	350–450	600–750	950–1,200
0.03	450–550	700–900	1,200–1,450
0.10($\hat{=} s_g$)	800–1,000	1,200–1,500	1,600–2,000

Note: Intermediate values can be interpolated linearly.

For bored piles with an enlarged base, the values must be reduced to 75%.

Table 4.5 Skin resistance $q_{s,k}$ for cohesive soils

Shear strength $c_{u,k}$ of the undrained soil [kN/m²]	Skin friction $q_{s,k}$ [kN/m²]
60	30–40
150	50–65
≥ 250	65–85

Note: Intermediate values can be interpolated linearly.

Table 4.6 Base resistance $q_{b,k}$ and skin resistance $q_{s,k}$ for rock

Confined compression strength $q_{u,k}$ [MN/m²]	Base resistance $q_{b,k}$ [kN/m²]	Skin friction $q_{s,k}$ [kN/m²]
0.5	1,500–2,500	70–250
5.0	5,000–10,000	500–1,000
20.0	10,000–20,000	500–2,000

Note: Intermediate values can be interpolated linearly.

To apply Tables 4.2 through 4.5, it must be assumed that

- $0.30 \text{ m} \leq D_s \leq 3.00 \text{ m}$
- Length into a stable soil layer is ≥ 2.50 m
- Thickness of the stable soil layer under the pile base is larger than $3 \cdot D_s$ resp. ≥ 1.50 m
- Stable soil layer has $q_c \geq 7.5$ MN/m² resp. $c_u \geq 100$ kN/m²

Independently in [2], it is recommended to put the pile toes in areas with $q_c \geq 10$ MN/m².

To apply Table 4.6 it must be assumed that

- Length in rock ≥ 0.50 m, if unconfined compression strength $q_u \geq 5$ MN/m²
- Length in rock ≥ 2.50 m, if unconfined compression strength $q_u \leq 0.5$ MN/m²
- Bedrock has a constant quality
- Spatial orientation of the rock surface as well as the spatial orientation of the gaps do not benefit a failure
- Gaps are not open and not filled with easily formable material
- Exclusion of a reduction of the strength as a result of a drilling process, for example, water in silt stone and clay stone

4.3.7 Pile load tests

The best opportunity for the determination of the bearing capacity of piles is a load test, which is performed *in situ* in the project area [32,33].

Pile load tests can be performed with a vertical or horizontal load for the determination of the vertical and the horizontal bearing capacity. In principle, load tests can be divided into static load tests and dynamic load tests. Detailed descriptions of the systems as well as information on the performance and the examination of the results are given in [2,9]. In the following, only the static pile load test for determining the vertical bearing capacity is presented.

For static pile load tests, counterweights or anchors are used. The installation of counterweights or anchors involves large technical and financial input. Using hydraulic jacks like the Osterberg-cell (O-cell), is more convenient. In a test pile, several hydraulic jacks can be installed to determine the skin friction in different pile segments that correspond to different soil layers. The single pile segments serve as counterweights for the different test phases. Figure 4.13 shows the variations of static pile load tests.

The result of a pile load test with vertical load is described by a resistance settlement curve $R_{c,k}(s)$, which can be used as the basis for the analyses of stability and serviceability. The pile resistance $R_{c,k}$, which is needed for the analysis of the limit state GEO-2, is determined out of the resistance settlement curve. Nevertheless, it is often difficult to determine the stability limit state by the resistance settlement curve because of its curve shape. Normally, the stability limit state resistance is defined at a maximum settlement of 10% of the pile base diameter ($s_g = 0.10\,D_b$). This is only applied if the value of the stability limit state can be identified explicitly, or if the stability limit state value appears at great settlements [9]. In other cases, other methods for the evaluation of the stability limit state can be used. Figure 4.14 shows an example of a qualitative trend of a resistance settlement curve. The pile resistance $R_{c,k}$ is determined by means of two straight lines. The first line draws a tangent on the beginning of the resistance settlement curve and the

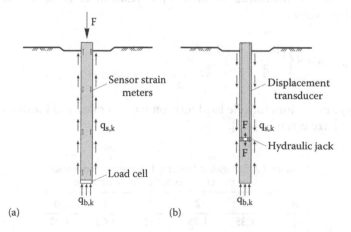

Figure 4.13 Static pile load test with counterweight resp. anchor (a) and a hydraulic jack (b).

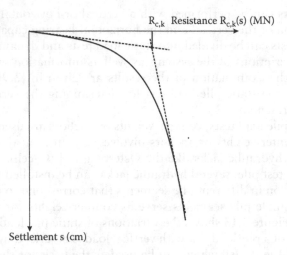

Figure 4.14 Determination of the pile resistance by a resistance settlement curve.

second line draws a tangent on the end of the resistance settlement curve. The intersection of both lines defines the stability limit state [34,35].

Based on one or several pile load tests, the measured value $R_{c,m}$ is determined, which has to be reduced by the factor ξ, taking straggling into account. For pressure piles where the superstructure is not able to transfer loads from softer to stiffer piles, the pile resistance has to be calculated by Equation 4.16. If the superstructure has a sufficient rigidity, it is able to transfer loads from softer to stiffer piles. In this case, the straggling factors ξ_i can be divided by 1.1, considering that ξ_1 is always ≥ 1.0. The straggling factor ξ_1 is for the measured average pile resistance and the straggling factor ξ_2 is for the measured minimum pile resistance. Table 4.7 shows the straggling factors ξ_i.

$$R_{c,k} = MIN\left\{\frac{(R_{c,m})_{av}}{\xi_1}; \frac{(R_{c,m})_{min}}{\xi_2}\right\} \qquad (4.16)$$

Straggling factors for pile load tests on tension piles and for dynamic pile load tests are given in [1,3].

Table 4.7 Straggling factor ξ_i for the determination of the characteristic pile resistance

n	1	2	3	4	≥5
ξ_1	1.35	1.25	1.15	1.05	1.00
ξ_2	1.35	1.15	1.00	1.00	1.00

Note: n is the number of the pile load tests.

4.3.8 Special methods for analysis

For defining the bearing capacity of piles, geotechnical analyses have been developed [9,36,37]. The developed theories of failure principally differ on the assumption of the failure body and the constitutive equations. Most theories are based on the simplification that piles are raft foundations that are very deeply embedded into the subsoil. The failure mechanism is described analogical to the base failure of classic raft foundations considering only the influence of the depth of the piles and the cohesion of the subsoil. The width of a pile is neglected due to the very small influence.

The assumption that pile foundations are like deep raft foundations leads to a bearing capacity that is increasing linearly with the depth. But several model tests show that this theoretical effect does not occur in reality [9].

Due to economic and/or technical reasons, for deep foundations with only a few piles and for offshore foundations, pile load tests are not applied. Therefore, theories are still developed to determine the bearing capacity of piles analytically.

4.3.9 Negative skin friction

The negative skin friction is counted among the special influences on pile foundations [38]. The negative skin friction results from an axial relative displacement between the subsoil and the pile, which is caused by settlements of a soft soil layer. Via skin friction the subsoil settlements stick on the pile. This skin friction is characterized as negative because it acts contrary to the skin friction of pile settlements. The settlements of soft soil layers result, for example, from additional loads, consolidation processes or a reduction of the groundwater level. For tension piles, negative skin friction occurs due to a heaving of the surrounding subsoil.

The characteristic value of the negative skin friction in cohesive subsoil can be evaluated by Equation 4.17. Thereby, the factor α is between 0.15 and 1.60, depending on the subsoil and the pile type. For an approximation, $\alpha = 1$ can be applied. The characteristic value of the negative skin friction in non-cohesive subsoil can be evaluated by Equation 4.18.

$$\tau_{n,k} = \alpha \cdot c_u \tag{4.17}$$

where:

$\tau_{n,k}$ = negative skin friction
c_u = shear strength of the undrained soil
α = factor for determining the size of negative skin friction of cohesive soils

$$\tau_{n,k} = \sigma'_v \cdot K_0 \cdot \tan\varphi' \tag{4.18}$$

where:

$\tau_{n,k}$ = negative skin friction
σ'_v = effective vertical stress
K_0 = coefficient of the earth pressure at rest
φ' = angle of friction

The negative skin friction $\tau_{n,k}$ is not larger than the positive skin friction q_s and reaches to the neutral point, from where only the positive skin friction takes effect.

If the bearing capacity of a pile foundation is adjusted by pile load tests, the influences of the negative skin friction have to be considered. Therefore, the measured positive skin friction in the compressible layer and in the layers above, in which the negative skin friction can occur, must be subtracted from the measured load value on the top of the pile [3].

4.3.10 Serviceability limit state (SLS)

The analysis of safety against loss of usability (serviceability limit state SLS) has to be focused on if the deformation of the pile foundation is significant for the rising construction. This is important with regard to absolute settlements as well as differential settlements.

The settlements of a pile group with a predominant load transfer via pile base pressure can be calculated analogical to a deep raft foundation. For a pile group with a predominant load transfer via skin friction, the settlement determinations are much more complex. If a precise determination of the deformation behavior of the pile foundation is necessary, it is recommended to apply a numerical simulation [2]. These numerical simulations have to be calibrated in an appropriate form, for example, on the basis of pile load tests [39–41].

A differential settlement can be determined as explained in Figure 4.15 [9]. The characteristic settlement s_{SLS} defines the permitted settlement of a

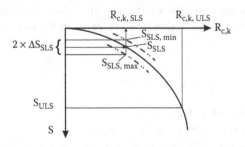

Figure 4.15 Determination of the expected differential settlement of a pile group.

building. To analyze potential impacts of differential settlements of neighboring piles, the settlement s_{SLS} has to be increased or reduced by a factor, so a range of $2\Delta s_{SLS}$ results. A cautious estimation is $\Delta s_{SLS} = 0.15 s_{SLS}$ if no further analyses are planned.

4.4 EXAMPLES OF CLASSIC PILE FOUNDATIONS FROM ENGINEERING PRACTICE

4.4.1 Commerzbank

The new high-rise building of the Commerzbank in Frankfurt am Main, Germany, is 259 m high; including the antenna, it is about 300 m high. Until 2003, it was the tallest building in Europe. It was constructed from 1994 to 1997, and it is founded on a classic pile foundation in rock [42]. Figure 4.16 gives an impression of the high-rise building. The ground view has a triangular shape with a length of each edge of about 60 m. The construction has three sublevels. The raft is 2.5 m to 4.45 m thick. The foundation depth of the raft is 12 m below the surface.

The settlement-relevant weight of the building is about 1400 MN [43]. The foundation consists of 111 bored piles with a length of 37.6 m to 45.6 m. The piles have a diameter of about 1.8 m in the Frankfurt Clay and a diameter of 1.5 m in the Frankfurt Limestone. The piles reach about 10 m into the rock material of the Frankfurt Limestone.

Figure 4.17 shows the measured settlements. The new Commerzbank high rise building has a settlement of 1.5 cm to 2 cm. The existing high-rise building, which is 103 m high, has additional settlements of up to 1.4 cm due to the new high-rise building.

4.4.2 PalaisQuartier

The challenging construction project "PalaisQuartier" has been realized in the center of Frankfurt am Main, Germany [44–46]. The project consists of two towers and adjacent buildings. The towers are up to 136 m high. Under the entire area, underground parking with five levels has been constructed. The foundation level is 22 m below the surface. The whole building complex has been constructed by using the top-down method. The structure is founded on 302 bored piles with a length of up to 27 m and a diameter of up to 1.86 m. The pile base is located in the stiff Frankfurt Limestone. Figures 4.18 and 4.19 give an impression of the whole building complex.

Due to the large number of piles an optimization of this classic deep foundation system was desired. Therefore, *in situ* pile load tests using Osterberg-cells (O-cells) were carried out. The bearing capacity of the piles in the Frankfurt Limestone with and without a shaft grouting was

Figure 4.16 High-rise building Commerzbank.

of importance. The test pile with a diameter of 1.7 m had two load cell levels between the three pile segments. The upper and the middle pile segment have a length of 5 m each. The lower segment has a length of 0.5 m (Figure 4.20). To increase the skin friction at the middle pile segment, a shaft grouting was carried out before starting the test. The upper and the middle pile segments were used for determination of the skin friction in the

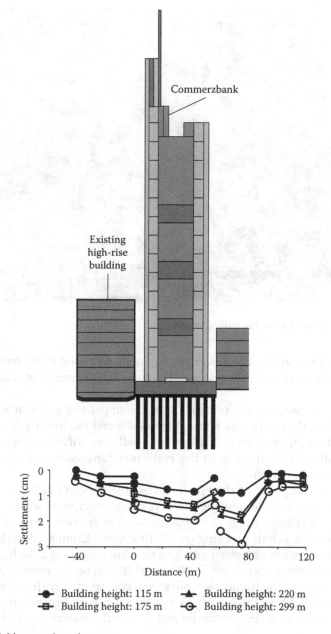

Figure 4.17 Measured settlements.

Figure 4.18 Building complex PalaisQuartier.

Frankfurt Limestone. The lower pile segment was used for determination of the base resistance at the pile toe. The rest of the borehole was filled with gravel.

In the first phase of the pile load test, the upper test segment was lifted by the upper O-cell. In this phase, the middle and the lower pile segments were used as an abutment. The lower O-cell was stiffened (Figure 4.21). The mobilized skin friction in the Frankfurt Limestone, without a shaft grouting was 830 kN/m².

In the second phase of the pile load test, the upper O-cell was released. Afterward, the middle and the lower pile segments were strained by the lower O-cell (Figure 4.22). The mobilized skin friction in the Frankfurt Limestone with a shaft grouting was 1040 kN/m². Compared to the results of test phase 1 without a shaft grouting the skin friction is 24% bigger. The pile base resistance was about 7000 kN/m². Based on the results of the pile load test, most of the foundation piles were constructed with a shaft grouting in the Frankfurt Limestone. A significant amount of pile meters and related construction time, material and financial resources could be saved.

4.4.3 International Business Centre Solomenka

In Kiev, Ukraine, a new International Business Centre Solomenka is under construction. The project consists of a 32-storey building, a shopping mall, an office building and underground parking. Figure 4.23 gives an

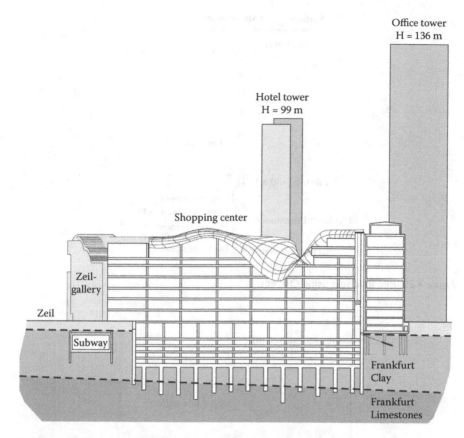

Figure 4.19 Cross-section through the building complex PalaisQuartier.

impression of the project. The foundation of the building complex was originally planned as a classic pile foundation with 167 bored piles with a diameter of 1.5 m and 99 barrettes with dimensions between 2.8 m×0.8 m and 6.8 m×0.8 m. The length of the planned deep foundation elements was 14 m to 46 m. The soil and groundwater conditions are shown in Figure 4.24.

Due to the large number of deep foundation elements, an optimization of this classic deep foundation system was necessary. Therefore, *in situ* pile load tests using O-cells were carried out at a sample pile with a diameter of 0.88 m and a length of 12 m. To investigate the pile base resistance in the clay and clay marl (Kiev formation), the pile base was constructed at a depth of 37 m under the ground surface. The empty boreholes were filled with gravel. Figure 4.25 shows the test setup of the pile. The test pile had two levels with O-cells (upper O-cell and lower O-cell). The upper O-cell was installed approximately 5.5 m above the pile toe. The lower O-cell was installed 50 cm above the pile to in the Kiev clay marl. The two O-cells

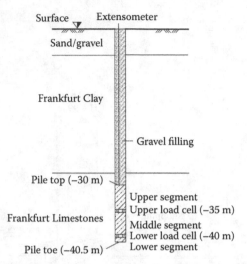

Figure 4.20 Pile load test using O-Cells.

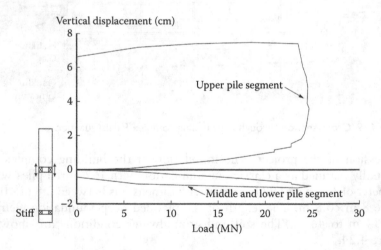

Figure 4.21 Test phase 1 for the determination of the skin friction of Frankfurt Limestone without a shaft grouting.

divide the test pile in an upper segment 1 with a length of 6.5 m, in a middle segment 2 with a length of 5 m, and in a lower segment 3 with a length of 0.5 m.

The test covered two different load phases. In the first phase, the upper O-cell was activated in order to determine the skin friction of the upper layers of the pile segment 1. The skin friction and the peak pressure of

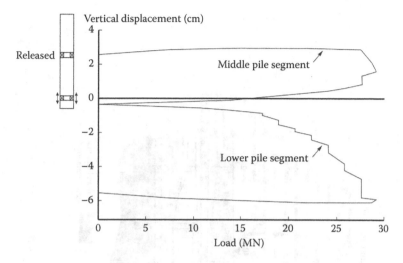

Figure 4.22 Test phase 2 for the determination of the skin friction of Frankfurt Limestone with a shaft grouting.

Figure 4.23 International Business Center Solomenka in Kiev, Ukraine.

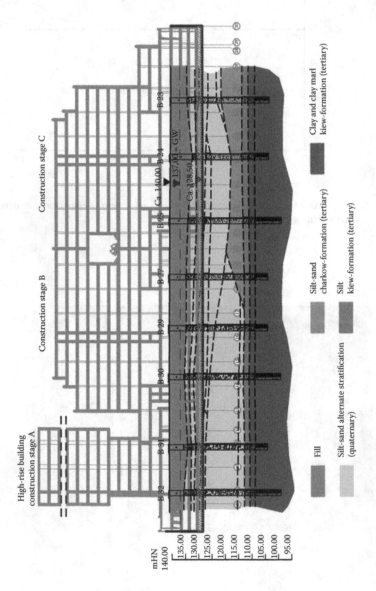

Figure 4.24 Subsoil and groundwater conditions.

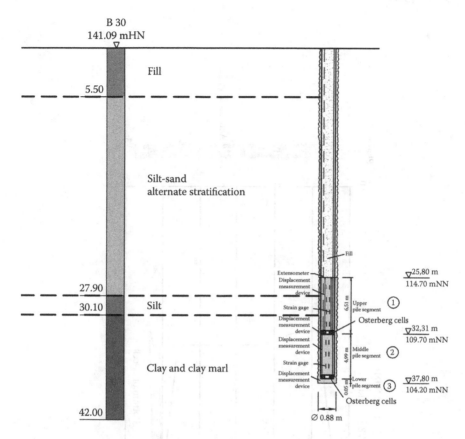

Figure 4.25 Test setup of the sample pile.

the Kiev clay marl are used as abutment. The applied maximum load was 1.2 MN. The load displacement curve of the first test phase is displayed in Figure 4.26. Due to the occurring displacements the ultimate limit state capacity of the pile segment 1 was reduced to 1 MN. The limit skin friction of the soil layers of pile segment 1 is about 65 kN/m².

The second test phase was split up into phase 2A and phase 2B. In the test phase 2A, the lower O-cell was activated, and the pile segment 2 was pressed upward. The upper O-cell was released, so that no load transmission to the pile segment 1 could take place. The load was increased up to the limit resistance of 1.15 MN of the pile segment 2. The load displacement curve of the test phase 2A and 2B is displayed in Figure 4.27. The limit skin friction of the soil layers of pile segment 2 in Kiev clay marl is about 80 kN/m². This is 25% higher than specified in Kiev up to this test. To determine the base resistance in Kiev clay marl in test phase 2B, the upper O-cell was stiffened and the lower O-cell was activated again. The load was increased up to the limit of the O-cells. The limit value of the base

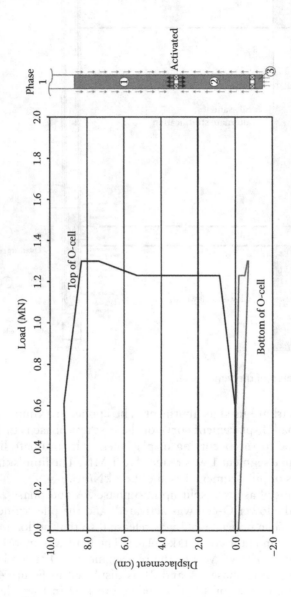

Figure 4.26 Results of test phase I.

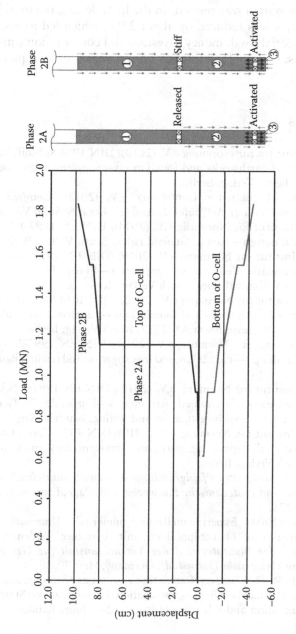

Figure 4.27 Results of test phase 2.

resistance of pile segment 3 could not be reached. The maximum mobilized base resistance is about 3000 kN/m^2.

Based on the results of the *in situ* pile load tests, the first planning of the deep foundation system was revised. In the final design, the total length of piles and barrettes was reduced by about 25%, which led to a significant reduction of material used, money invested, and construction time needed. By using O-cells, the pile load test has been carried out very precisely and efficiently.

REFERENCES

1. Deutsches Institut für Normung e.V. (2010): DIN 1054 Subsoil: Verification of the safety of earthworks and foundations—Supplementary rules to DIN EN 1997-1. Beuth Verlag, Berlin.
2. Deutsche Gesellschaft für Geotechnik e.V. (2012): *Empfehlungen des Arbeitskreises Pfähle (EA-Pfähle)*. 2. Auflage, Ernst & Sohn Verlag, Berlin.
3. Deutsches Institut für Normung e.V. (2014): DIN EN 1997-1 Eurocode 7: Geotechnical design—Part 1: General rules. Beuth Verlag, Berlin.
4. Deutsches Institut für Normung e.V. (2010): DIN EN 1997-1/NA National Annex: Nationally determined parameters—Eurocode 7: Geotechnical design—Part 1: General rules. Beuth Verlag, Berlin.
5. Deutsches Institut für Normung e.V. (2012): DIN 1054 Subsoil: Verification of the safety of earthworks and foundations—Supplementary rules to DIN EN 1997-1:2010; Amendment A1: 2012. Beuth Verlag, Berlin.
6. Deutsches Institut für Normung e.V. (2010): DIN EN 1997-2 Eurocode 7: Geotechnical design—Part 2: Ground investigation and testing. Beuth Verlag, Berlin.
7. Deutsches Institut für Normung e.V. (2010): DIN EN 1997-2/NA National Annex: Nationally determined parameters—Eurocode 7: Geotechnical design—Part 2: Ground investigation and testing. Beuth Verlag, Berlin.
8. Deutsches Institut für Normung e.V. (2010): DIN 4020 Geotechnical investigations for civil engineering purposes: Supplementary rules to DIN EN 1997-2. Beuth Verlag, Berlin.
9. Kempfert, H.-G. (2009): *Pfahlgründungen. Grundbautaschenbuch, Teil 3: Gründungen und geotechnische Bauwerke*. 7. Auflage, Ernst & Sohn Verlag, Berlin, 73–278.
10. Schmitt, A. (2004): Experimentelle und numerische Untersuchungen zum Tragverhalten von Ortbetonpfählen mit variabler Bodenverdrängung. *Mitteilungen des Institutes und der Versuchsanstalt für Geotechnik der Technischen Universität Darmstadt*, Germany, Heft 70.
11. Katzenbach, R. (2013): Deep foundation systems as economic and save solutions for geotechnical challenges. International Conference on State of the Art of Pile Foundation and Pile Case Histories, 2–4 June, Bandung, Indonesia, VII–IX.
12. Hudelmaier, K.; Küfner, H. (2008): *Spezialtiefbau Kompendium: Verfahrenstechnik und Geräteauswahl*. Ernst & Sohn Verlag, Berlin.

13. Hudelmaier, K.; Küfner, H. (2009): *Spezialtiefbau Kompendium Band II: Verfahrenstechnik und Geräteauswahl.* Ernst & Sohn Verlag, Berlin.
14. Eds. Katzenbach, R.; Leppla, S. (2015): *Handbuch des Spezialtiefbaus.* 3. Auflage, Bundesanzeiger Verlag, Cologne, Germany.
15. Deutsches Institut für Normung e.V. (2010): DIN EN 1536 Execution of special geotechnical work: Bored piles; German version EN 1536:2010. Beuth Verlag, Berlin.
16. Deutsches Institut für Normung e.V. (2012): DIN SPEC 18140 Supplementary provisions to DIN EN 1536:2010-12, Execution of special geotechnical works: Bored piles. Beuth Verlag, Berlin.
17. Deutsches Institut für Normung e.V. (2007): DIN EN 12794 Precast concrete products: Foundation piles; German version EN 12794:2005+A1:2007. Beuth Verlag, Berlin.
18. Deutsches Institut für Normung e.V. (2001): DIN EN 12699 Execution of special geotechnical work: Displacement piles; German version EN 12699:2000. Beuth Verlag, Berlin.
19. Deutsches Institut für Normung e.V. (2010): DIN EN 12699 Execution of special geotechnical works: Displacement piles; German version EN 12699:2000, Corrigendum to DIN EN 12699:2001-05. Beuth Verlag, Berlin.
20. Deutsches Institut für Normung e.V. (2012): DIN SPEC 18538 Supplementary provisions to DIN EN 12699:2005-05, Execution of special geotechnical work - Displacement piles. Beuth Verlag, Berlin.
21. Deutsches Institut für Normung e.V. (2012): DIN EN 14199 Execution of special geotechnical works: Micropiles; German version EN 14199:2005. Beuth Verlag, Berlin.
22. Deutsches Institut für Normung e.V. (2012): DIN SPEC 18539 Supplementary provisions to DIN EN 14199:2012-01, Execution of special geotechnical works: Micropiles. Beuth Verlag, Berlin.
23. Deutsches Institut für Normung e.V. (1999): DIN EN 12063 Execution of special geotechnical works: Sheet-pile walls; German version EN 12063:1999. Beuth Verlag, Berlin.
24. Deutsches Institut für Normung e.V. (2012): DIN 18301 German construction contract procedures (VOB): Part C: General technical specifications in construction contracts (ATV): Drilling works. Beuth Verlag, Berlin.
25. Deutsches Institut für Normung e.V. (2012): DIN 18304 German construction contract procedures (VOB): Part C: General technical specifications in construction contracts (ATV)—Piling. Beuth Verlag, Berlin.
26. Deutsches Institut für Normung e.V. (2013): DIN 4126 Stability analysis of diaphragm walls. Beuth Verlag, Berlin.
27. Deutsches Institut für Normung e.V. (2013): DIN 4126 Stability analysis of diaphragm walls: Supplement 1: Explanations. Beuth Verlag, Berlin.
28. Deutsches Institut für Normung e.V. (2014): DIN 4127 Earthworks and foundation engineering: Test methods for supporting fluids used in the construction of diaphragm walls and their constituent products. Beuth Verlag, Berlin.
29. Deutsches Institut für Normung e.V. (2015): DIN EN 1538 Execution of special geotechnical work: Diaphragm walls (includes Amendment A1:2015). Beuth Verlag, Berlin.

30. Deutsches Institut für Normung e.V. (2012): DIN 18313 VOB Vergabe- und Vertragsordnung für Bauleistungen—Teil C: Allgemeine Technische Vertragsbedingungen für Bauleistungen (ATV): Schlitzwandarbeiten mit stützenden Flüssigkeiten. Beuth Verlag, Berlin.
31. Pulsfort, M. (2012): Grundbau, Baugruben und Gründungen. Handbuch für Bauingenieure: Technik, Organisation und Wirtschaftlichkeit, Springer-Verlag, Heidelberg, Germany, 1568–1639.
32. Briaud, J.-L.; Ballouz, M.; Nasr, G. (2000): Static capacity prediction by dynamic methods for three bored piles. *Journal of Geotechical and Geoenvironmental Engineering*, Vol. 126, July, ASCE, Reston, VA, 640–649.
33. Katzenbach, R. (2005): Optimised design of high-rise building foundations in settlement-sensitive soils. *International Geotechnical Conference of Soil–Structure-Interaction*, 26–28 May, St. Petersburg, Russia, 39–46.
34. Fuller, R.M.; Hoy, H.E. (1970): Pile load tests including quick-load test method, conventional methods and interpretations. *Research Record HRB 333, Highway Research Board*, Washington, D.C., USA, 1970, 74–86.
35. Butler, H.D.; Hoy, H.E. (1977): The Texas Quick-Load Method for Foundation Load Testing: User's Manual. Report No. FHWA-IP-77-8, Federal Highway Administration, Office of Development, Washington, D.C., 59 S.
36. Skempton, A.W. (1951): The bearing capacity of clays. *Building Research Congress*, September, London, UK, 180–189.
37. Kolymbas, D. (1989): *Pfahlgründungen*. Springer-Verlag, Heidelberg, Germany.
38. Kempfert, H.-G. (2005): Negative Mantelreibung bei Pfahlgründungen nach dem Teilsicherheitskonzept. *Mitteilungen des Instituts und der Versuchsanstalt für Geotechnik der Technischen Universität Darmstadt*, Germany, Heft 71, 21–31.
39. Katzenbach, R.; Leppla, S. (2014): Deep foundation systems for high-rise buildings in difficult soil conditions. *Geotechnical Engineering Journal of the SEAGS & AGSSEA*, Vol. 45, No. 2, June, 115–123.
40. Katzenbach, R.; Leppla, S.; Krajewski, W. (2014): Numerical analysis and verification of the soil–structure interaction in the course of large construction projects in inner cities. *International Conference on Soil–Structure Interaction: Underground Structures and Retaining Walls*, 16–18 June, St. Petersburg, Russia, 28–34.
41. Katzenbach, R., Leppla, S. (2015): Optimised and safety of urban geotechnical construction projects. *16th African Regional Conference on Soil Mechanics and Geotechnical Engineering*, 27–30 April, Hammamet, Tunisia, 233–239.
42. Katzenbach, R.; Hoffmann, H.; Vogler, M.; Moormann, C. (2001): Cost-optimized foundation systems of high-rise structures, based on the results of actual geotechnical research. *International Conference Trends in Tall Buildings*, 5–7 September, Frankfurt am Main, Germany, 421–443.
43. Holzhäuser, J. (1998): Experimentelle und numerische Untersuchungen zum Tragverhalten von Pfahlgründungen im Fels. *Mitteilungen des Instituts und der Versuchsanstalt für Geotechnik der Technischen Universität Darmstadt*, Germany, Heft 42.

44. Janke, O.; Zoll, V.; Sommer, F.; Waberseck, T. (2010): PalaisQuartier (FrankfurtHochVier)—Herausfordernde Deckelbauweise im Herzen der City. *Mitteilungen des Instituts und der Versuchsanstalt für Geotechnik der Technischen Universität Darmstadt*, Germany, Heft 86, 113–124.

45. Vogler, M. (2010): Berücksichtigung innerstädtischer Randbedingungen beim Entwurf tiefer Baugruben und Hochhausgründungen am Beispiel des PalaisQuartier in Frankfurt am Main. *Bauingenieur 85*, June, Springer VDI Verlag, Düsseldorf, Germany, 273–281.

46. Katzenbach, R.; Leppla, S.; Waberseck, T. (2012): Deep excavations and deep foundation systems combined with energy piles. *Baltic Piling Days*, 3–5. September, Tallinn, Estonia, 14 S.

Chapter 5

Combined pile-raft foundation (CPRF)

A CPRF is a special form of deep foundation. A CPRF is a hybrid foundation system that combines the bearing capacity of a foundation raft and the piles or barrettes. A CPRF is a technically and economically optimized foundation system. CPRFs can be used for the foundations of classic high-rise buildings as well as for engineering constructions like bridges and towers.

For CPRFs, the same technical regulations as for classic deep foundations can be applied [1]. An additional regulation is the Combined Pile-Raft Foundation Guideline [2], which reflects the individual features of a CPRF. The CPRF Guideline is internationally valid and published by the International Society for Soil Mechanics and Geotechnical Engineering (ISSMGE) [3].

Due to the interaction between the foundation elements and the subsoil, CPRFs have a very complex bearing and deformation behavior. Therefore, CPRFs have to be classified into the Geotechnical Category 3 according to EC 7. For safety and quality assurance, not only does an independent checking engineer have to be involved for structural design, but also an independent checking engineer of geotechnics, has to be involved to guarantee the four-eye principle [2].

The advantages of a CPRF, compared to a conventional spread foundation and a classic pile foundation, can be summarized as follows:

- Reduction of settlements and differential settlements
- Increasing of the bearing capacity of spread foundations
- Reduction of the bending moments of the foundation raft
- Reduction of pile materials (30%–50%) [4]

5.1 BEARING AND DEFORMATION BEHAVIOR

Measurement data of high-rise buildings with spread foundations in Frankfurt am Main, Germany, showed the load transfer into the subsoil. Between 60% and 80% of the settlements arise in the upper third of the

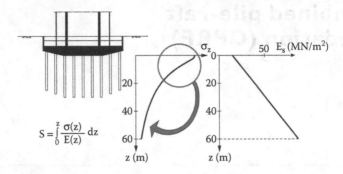

$$S = \int_0^z \frac{\sigma(z)}{E(z)}\, dz$$

Figure 5.1 Load transfer of a CPRF.

influenced soil volume (Figure 5.1) [5]. A CPRF transfers a part of the stresses from areas with a small stiffness under the foundation raft to a stiffer, deeper area of the subsoil. This transmission is effected via the piles of the CPRF without neglecting the bearing capacity of the foundation raft.

According to [1], a CPRF can be constituted as a geotechnical composite structure, which consists of the following, interacting bearing elements:

- Piles
- Foundation raft
- Subsoil

The bearing and deformation behavior is characterized by the interaction between the bearing elements and the subsoil. Figure 5.2 shows all interactions of a CPRF. Due to the stiffness of the foundation raft, the loads $F_{tot,k}$ of the rising structure are transferred to the piles and the subsoil. Similar to a classic deep foundation, the mobilized resistance of a CPRF depends significantly on the settlement s. The integration of the soil contact pressure $\sigma(x,y)$ under the foundation raft is the resistance $R_{raft,k}(s)$. The resistance of the foundation piles $\Sigma R_{pile,k}(s)$ added to the resistance of the foundation raft $R_{raft,k}(s)$ delivers the total resistance $R_{tot,k}(s)$ of a CPRF (Equation 5.1). With Equation 5.2, the resistance of a single foundation pile i can be determined. The resistance of a single pile of a CPRF consists of the skin resistance $R_{s,k,i}(s)$ and the pile base resistance $R_{b,k,i}(s)$. The skin resistance $R_{s,k,i}(s)$ can be calculated by integration of the skin friction $q_{s,k}(s,z)$, which depends on the settlement s and the depth z.

$$R_{tot,k}(s) = \sum_{i=0}^{n} R_{pile,k,i}(s) + R_{raft,k}(s) \tag{5.1}$$

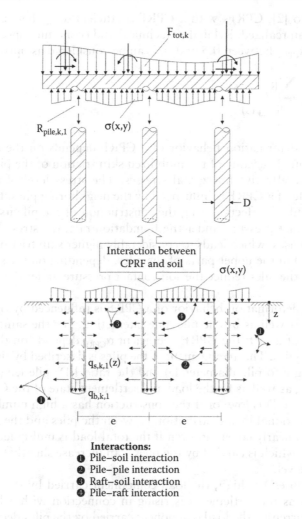

Interactions:
❶ Pile–soil interaction
❷ Pile–pile interaction
❸ Raft–soil interaction
❹ Pile–raft interaction

Figure 5.2 Interactions of the CPRF.

$$R_{pile,k,i}(s) = R_{b,k,i}(s) + R_{s,k,i}(s)$$

$$= q_{b,k,i} \cdot \frac{\pi \cdot D^2}{4} + \int q_{s,k,i}(s,z) \cdot \pi \cdot D \cdot dz \qquad (5.2)$$

The load deformation behavior of a CPRF can be specified by the CPRF coefficient α_{CPRF}. The coefficient declares the relation between the resistance of the piles and the total resistance (Equation 5.3). The CPRF coefficient varies between 0 and 1. A CPRF coefficient $\alpha_{CPRF} = 0$ means that the whole load $F_{tot,k}$ is carried by the foundation raft. A CPRF coefficient $\alpha_{CPRF} = 1$ means that the whole load $F_{tot,k}$ is carried by the foundation piles.

According to [2], CPRFs with a CPRF coefficient α_{CPRF} between 0.3 and 0.9 have been realized. Related to technical and economic aspects, a CPRF coefficient α_{CPRF} between 0.5 and 0.7 can be considered as optimum.

$$\alpha_{CPRF} = \frac{\sum R_{pile,k,i}(s)}{R_{tot,k}(s)} \tag{5.3}$$

The load deformation behavior of a CPRF depends on the stress state of the subsoil. The value of the mobilized skin friction of the piles is influenced by the effective horizontal stresses. The stress level of the subsoil around a pile of a CPRF is influenced by the neighboring piles, the foundation raft, and the effects during the construction of the pile itself. Due to the soil contact pressure under the foundation raft, the stress level of the subsoil increases, which leads to considerably higher skin frictions that can be mobilized in the upper parts of the piles, depending on the settlements. Conversely, the piles reduce the soil contact pressure under the foundation raft.

The load deformation behavior of a CPRF is influenced by the pile–raft interaction as well as by the pile–pile interaction. For the same geometric dimensions of a raft, the CPRF coefficient α_{CPRF} depends on the arrangement of the piles. The arrangement of the piles is described by the ratio e/D (pile spacing e to pile diameter D) and the ratio l/D (pile length l to pile diameter D), as well as by the load and settlement state of the CPRF.

If the ratio e/D is low, or if the construction has a high number of piles n, the proportional load distribution between the piles and the foundation raft remains nearly constant, even if the total load is multiplied. The load component, which is carried by the piles, can increase slightly because of a rising load level.

If the ratio e/D is high, the load component carried by the foundation raft as well as the settlements is rising in connection with a higher load level. Concurrently, the load component carried by the piles decreases. The reason for this is the different load-bearing behavior of the piles, which is influenced by the pile spacing e.

If the pile spacing e between the piles of a CPRF decreases, the deviation of the load bearing of a single pile increases. If the ratio $e/D \geq 3$, the load-bearing behavior of a pile of a CPRF essentially depends on the position within the foundation system.

Analogically to classic pile foundations, at the same settlements the pile resistance of a CPRF increases from the center to the edge of the raft. Due to the neighboring piles, the inner piles receive significantly smaller loads. The edge piles, especially the corner piles, receive the majority of the loads.

Due to the missing shielding by neighboring piles, the edge piles and corner piles offer a significantly higher rigidity. The pile resistance, which depends on the pile position, is mainly caused by the different skin resistances, while the pile base resistance is nearly independent from the pile position.

With growing pile spacing e between the piles, at a CPRF the influence of the neighboring piles gets smaller, analogue to a classic pile foundation. If the ratio $e/D \geq 6$, all piles of a CPRF have the same load deformation behavior independent of their position. There is no group effect.

At a classic pile foundation, the mobilized skin friction at the upper parts of the piles is comparatively small, which can be explained by the low stress level. In contrast, larger settlements at a CPRF lead to bigger skin frictions in the upper parts of the piles, which can be explained by the higher stress level.

5.2 CALCULATION METHODS

For the design and calculation of a CPRF, various calculation methods can be selected, all of them based on different calculation methods and modeling schemes. Detailed descriptions and literature references to the different methods can be found in [2,6–13].

The results of the different methods depend on the modeling schemes and the simplified assumptions and boundary conditions. Only for a preliminary design or in very simple cases are these methods sufficient to develop a technically and economically optimized CPRF. Only the numerical methods provide calculation results that are comparable to the reality.

The following calculation methods are available [2].

- Empirical methods: Based on the results of laboratory and field tests, the bearing capacity of a pile can be determined via correlations and the use of tabular values. Using further correlations, the bearing capacity of a pile group is determined. The correlations and empirical equations are based on experiences made at *in situ* measurements and model tests.
- Methods with equivalent alternative models: The CPRF is seen as an alternative model, such as a deep spread foundation or a thick single pile.
- Analytic methods: For example, at first, the bearing capacity of the foundation raft is determined, neglecting the piles. If the loads are higher than the bearing capacity of the raft, the surplus loads are allocated to the piles. The piles are regarded as single piles whose resistance can be fully activated. The settlements are determined for the foundation raft and its respective load. Due to its disregard of all interactions, this approach is only convenient as a first assessment.
- Numerical methods: The finite element method (FEM) is the generally used numerical method, being able to consider and solve complicated geometries and nonlinear constitutive equations. In three-dimensional

simulations, the foundation elements are modeled with linear elastic material behavior, while the subsoil is modeled with elastoplastic material behavior.

5.3 GEOTECHNICAL ANALYSIS

5.3.1 Ultimate limit state (ULS)

Analogous to the classic pile foundations, CPRFs have to be analyzed in accordance to the internal and external bearing capacity. The analysis of the internal bearing capacity is conducted according to the corresponding standards and regulations. The analysis of the external bearing capacity is conducted under consideration of the time-dependent material behavior of the subsoil and the rigidity of the rising structure.

According to [14], the external bearing capacity (geotechnical analysis GEO-2) is sufficient, if the design value of the loads E_d is smaller or equal to the design resistance $R_{tot,d}(s)$ ($E_d \leq R_{tot,d}(s)$). The design value of the resistance $R_{tot,d}(s)$ is defined by Equation 5.4. The total resistance of a CPRF is calculated for the whole foundation system, which consists of the foundation raft and the piles. A separate geotechnical analysis of the single piles is not necessary.

$$R_{tot,d}(s) = \frac{R_{tot,k}(s)}{\gamma_{R,v}} \tag{5.4}$$

The characteristic resistance must be calculated by using a validated model, for example, numerical methods. In simple cases, the application of other methods, for example, analysis of the base failure, is possible. Criteria of simple cases are

- Simple geometry (similar pile length and diameter; constant pile spacing; rectangular, quadratic, or round foundation raft; excess length of the foundation raft over the outer piles $\leq 3D_s$)
- Homogenous subsoil (no major differences of the stiffness)
- Centric loads
- No predominating dynamic loads

If for the analysis the base failure method is used, the resistance is specified by the lower edge of the foundation raft. The vertical bearing capacity of the piles is negligible. Indeed, the anchor bolt resistance of the piles in the failure surface can be considered [15].

The basics of the analysis of the ULS of a CPRF are summarized in the "ISSMGE Combined Pile-Raft Foundation Guideline" [3], which is fully printed in the appendix of this book.

5.3.2 Serviceability limit state (SLS)

For the assessment of the suitability of a CPRF, a maximum settlement, resp. a maximum settlement difference, must be specified. In the context of the verification of the serviceability, the specified limit values must be proved under characteristic loads. The specified limit values for the deformation are defined by the rising structure as well as by neighboring buildings and structures on the surface and under the surface.

The analysis of the internal bearing capacity of SLS is conducted according to the corresponding standards and regulations. This includes, for example, the limitation of crack widths of reinforced-concrete components.

5.3.3 Pile load tests

5.3.3.1 Basics

According to [1], for the design and the calculation of a CPRF, knowledge about the load deformation behavior of a free, single pile is necessary. If there is no knowledge of the bearing capacity of piles for a special pile type and in comparable soil conditions, a pile load test has to be performed. If a pile load test is not performed, the bearing capacity of a single pile can be estimated on the basis of empirical values, whereby this simplification and transferability has to be proved.

Knowledge about the bearing capacity of a free, single pile is important for two reasons. On the one hand, it is the only way to evaluate the selected geometries of the piles in accordance to the technical and environmental aspects and to prove the plausibility of the calculation method. On the other hand, it is possible to calibrate the numerical models. For complex construction projects and/or difficult soil conditions, *in situ* pile load tests are strongly advised.

5.3.3.2 Examples

For the new construction of a high-rise building in soft subsoil, numerical simulations have been carried out for the design of a CPRF [16]. The calibration of the numerical simulations is based on the results of a pile load test, which has been carried out on the project area. For the loading mechanisms, Osterberg-cells (O-cells) were used. The test pile consisted of three segments: The upper test segment 1, the middle test segment 2 between the two O-cells, and the lower test segment 3.

For the determination of the pile base resistance and the skin friction of the different soil layers, the individual O-cells were activated in various testing phases. For the determination of the skin friction and the pile base resistance of the test segment 3, only the lowest O-cell was activated, while test segment 2 was used as an abutment. For the determination of the skin friction of the test segment 2, the upper O-cell was activated, while the lower O-cell was released. In this test phase, test segment 1 was used as the abutment. For the determination of the skin friction of test segment 1, the upper O-cell was activated and the lower O-cell was stiffened. During this test phase, the test segments 2 and 3 were used as abutments.

The calibration of the numerical simulations of the CPRF was determined by a numerical back analysis of the pile load test using FEM. Figure 5.3 illustrates the setup of the test and the mesh of the FEM simulation with the three test segments and the two O-cells. The results of the *in situ* pile load test and the numerical back analysis are illustrated in Figure 5.4. The displayed curves show a good accordance. Based on the results of the numerical back analysis, the determined soil mechanical parameters were adjusted. Moreover, the simplified stratigraphy, which was necessary for the numerical simulations, was verified.

The design of the CPRF is performed by three-dimensional, nonlinear FE-simulations. The length, diameter and the number of the piles were optimized on the basis of the FE-simulations taking into account the

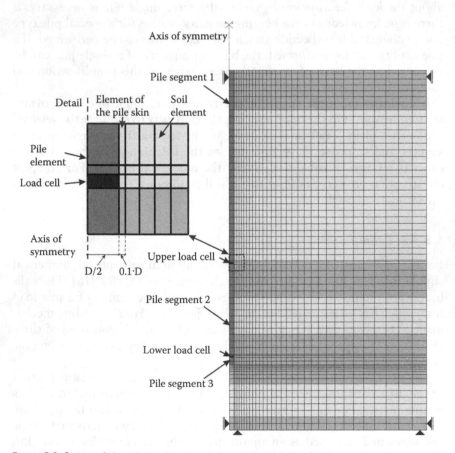

Figure 5.3 Setup of the pile load test and the numerical simulation.

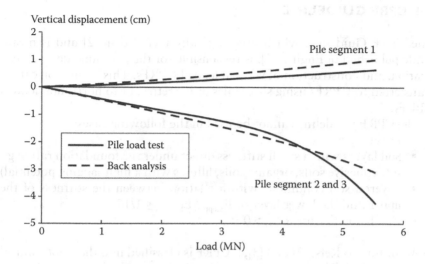

Figure 5.4 Results of the *in situ* pile load test and the numerical simulation (upper O-cell activated, lower O-cell stiffened).

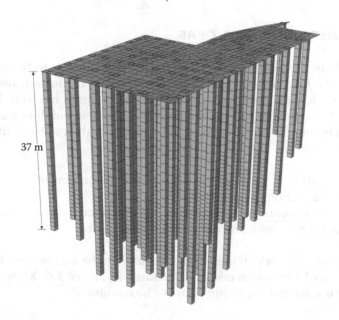

Figure 5.5 FE-Mesh of the optimized CPRF.

requirements of the load deformation behavior. Figure 5.5 illustrates the optimized CPRF in the FE-simulation. The CPRF coefficient is $\alpha_{CPRF} = 0.8$, which means that 80% of the total building weight is carried by the piles and 20% of the total buildings weight is carried by the raft.

5.4 CPRF GUIDELINE

The CPRF Guideline, which was originally included in [2] and is meanwhile published in English [3], is responsible for the planning, design, verification, and construction of vertical loaded CPRFs. This regulation can be transferred to CPRFs using sheet piles or barrettes, or to horizontal loaded CPRFs.

The CPRF Guideline cannot be used in the following cases:

- Soil layers with a small stiffness direct under the foundation raft (e.g., soft cohesive soils, organic soils, fillings with a high sagging potential)
- Layered soil conditions with a relation between the stiffness of the upper and the lower layer of $E_{S,upper}/E_{S,lower} \leq 1/10$
- CPRF coefficients $\alpha_{CPRF} > 0.9$

According to Refs. [2] and [3], a CPRF is classified into the Geotechnical Category GC 3. Therefore, the requirements are very high on the soil investigation, the construction, and the supervision, including the measurements.

5.5 MONITORING OF A CPRF

Very early in a planning stage a monitoring program has to be developed to check the load deformation behavior as well as the force transmission within a CPRF [1]. Therefore, the monitoring program has to include the geodetic and geotechnical measurements of the new building and of the vicinity. During the building phase and the serviceability phase, the monitoring program has to fulfill the following tasks:

- Verification of the calculation model and the parameters used
- Early detection of critical states
- Construction-related verification of the predicted deformations
- Quality assurance and preservation of evidence

The parameters, which need to be measured in the area of the CPRF and the associated measurements, are illustrated in Figure 5.6. Moreover, the construction pit and the vicinity have to be monitored.

5.6 EXAMPLES FROM ENGINEERING PRACTICE

A CPRF is a technically and economically optimized foundation system, which is established in engineering practice, but it is still an open research topic. Numerous examples show its excellent applicability [4,17,18]. The scope of application includes not only vertical loaded CPRFs, but also

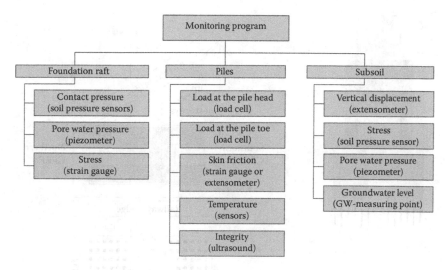

Figure 5.6 Monitoring of a CPRF.

horizontal loaded CPRFs [19]. The following subsections demonstrate several construction projects where CPRFs were applied to realize an economical and reliable basis for high-rise structures.

5.6.1 Messe Torhaus

The construction of the Messe Torhaus office building in 1983–1985 was the first application for a CPRF in Frankfurt am Main, or indeed anywhere in Germany [20]. The building has up to 30 storeys (Figure 5.7). Due to an adjacent triangular intersection of railway bridges, a settlement-restricted foundation was required. The office building is founded on two separate rafts and is underpassed by a road. The subsoil consists of quaternary gravel down to 5.5 m below the surface, underlain by Frankfurt Clay extending to a great depth.

The CPRF consists of two separate rafts, each with 42 bored piles with a length of 20 m and a diameter D of 0.9 m. The 6×7 piles of each raft are arranged symmetrically with a pile spacing e of 3–3.5 times of the pile diameter D. Both rafts have dimensions of 17.5 m×24.5 m and are founded 3 m below the surface. Each raft carries an effective structural load of 200 MN.

As there had been no precedent, the design of the CPRF for the Messe Torhaus was based on a conventional approach recommended by the German codes for fully piled foundations. Here the assumption was that all piles would be utilized at their ultimate bearing capacity, as known for a single pile, with the remaining part of the structural load transmitted directly by the raft to the subsoil.

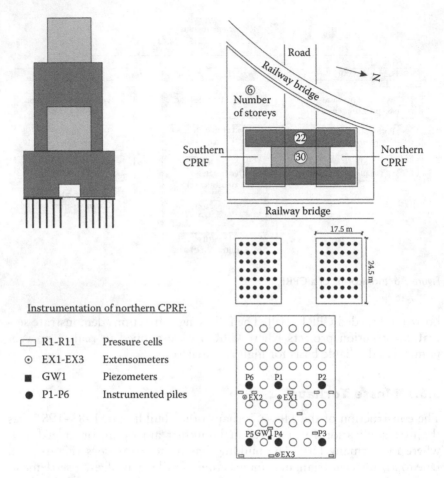

Figure 5.7 Messe Torhaus.

During construction, the behavior of the CPRF was monitored by a geodetic and a geotechnical measurement program. As shown in Figure 5.7 for the northern CPRF, 6 piles have been instrumented with strain gauges and a load cell at the pile base, 11 earth pressure cells have been installed beneath the raft, and 3 extensometers lead down to a depth of 40.5 m below the raft [20].

Figure 5.8 illustrates the measured load-settlement behavior of the northern CPRF, dividing the total structural load (R_{tot}) into the load carried by the raft (R_{raft}) and by the piles ($\Sigma R_{pile,i}$). The measurements indicate that only a small part of the structural load is transferred by the raft to the soil. The time–load and time–settlement curves in Figure 5.9 show that from the early beginning of concreting the raft, the contact pressure between raft and soil mainly comprises the dead weight of the raft, as

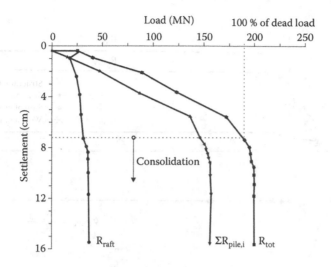

Figure 5.8 Measured load–settlement curves.

the additional loads of the superstructure are transferred by the piles to the subsoil [21]. The CPRF coefficient α_{CPRF} is 0.8. Due to the fast construction process, 95% of the dead load of the building had been applied within 8 months of the start of the construction. At the same time, only 40% of the final settlement had been recorded. The consolidation of the Frankfurt Clay led to additional settlements that were still continuing 3 years after the completion of the superstructure. During the consolidation process, the load distribution between raft and piles has remained constant.

The settlement distribution with depth measured by the extensometers is shown in Figure 5.10 for the center (EX1) and the edge (EX3) of the northern CPRF. As the extensometers were installed after concreting the raft, only additional settlements caused by the load of the superstructure were recorded [20]. Down to the level of the pile bases, the settlements remain approximately constant, indicating a block deformation of the piles and the surrounding soil.

The different pile loads are a consequence of the dependency of the mobilized skin friction, depending on the position of a pile within the group. Figure 5.11 shows the measured distributions of pile force and skin resistance for the corner pile and the inner pile. The corner pile mobilized an average skin friction of 140 kN/m² in the lower two-thirds of the pile shaft, whereas the inner pile mobilized a skin friction of only 60 kN/m² in the lowest third of the pile shaft. The value of 140 kN/m² for the skin friction of the corner pile is more than twice the value of the ultimate skin friction of 60 kN/m² determined from static load tests on short, single piles in Frankfurt Clay [22].

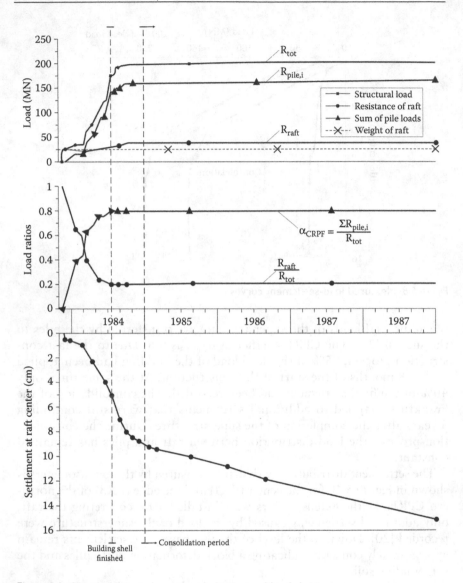

Figure 5.9 Observed time-dependent behavior and load sharing.

5.6.2 Messeturm

The Messeturm in Frankfurt am Main, Germany, is 256.5 m high. It is founded on a CPRF in the Frankfurt Clay (Figure 5.12). The foundation raft has a ground view of 58.8 m × 58.8 m with a maximum thickness of 6 m in the center and a thickness of 3 m at the edges. The base of the foundation raft is about 11–14 m below the surface. The raft is combined with

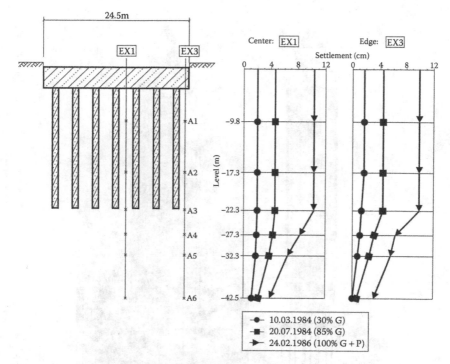

Figure 5.10 Measured settlement distribution.

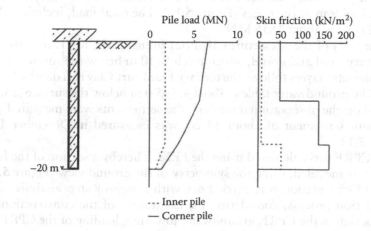

Figure 5.11 Measured distributions of pile load and skin friction.

Frankfurt
Clay

Figure 5.12 Messeturm.

64 bored piles with a diameter of 1.3 m and a length of 30.9 m in the center ring and 26.9 m at the edges (Figure 5.13). The total load, including 30% of live loads, is about 1855 MN.

In the area of the Messeturm artificial filling at the top is underlain by quaternary sand and gravel, which reach 8–10 m below the surface. Below the quaternary layers follows the tertiary Frankfurt Clay to a depth of about 70 m. The groundwater table is about 4.5–5.0 m below the surface [23,24].

Based on the observational method, the settlements were measured. The maximum settlement of about 13 cm was measured in December 1998 (Figure 5.14).

The CPRF was calculated using the FEM. Thereby a section of the foundation was modeled, using the symmetry of the ground view (Figure 5.15).

The FE-calculation was carried out with a step-by-step analysis of the construction process, considering the excavation of the construction pit, construction of the CPRF, groundwater lowering, loading of the CPRF and groundwater re-increase.

Single calculations were carried out with the FE model for different foundation systems. Thereby, different pile configurations and pile length were

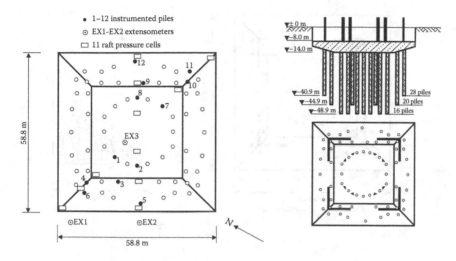

Figure 5.13 Ground view and cross-section of the CPRF.

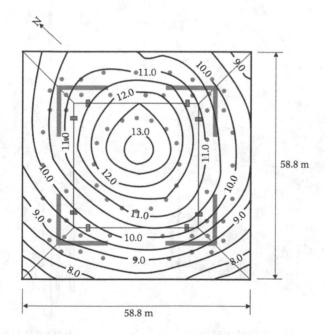

Figure 5.14 Settlements (cm) measured in December 1998.

analyzed, as well as a pure raft foundation. In Figure 5.16, the calculated settlements of the realized CPRF are compared with the settlements, calculated for a pure raft foundation.

The maximum settlements of a pure raft foundation were calculated to 32.5 cm. The calculated maximum settlements of the CPRF are nearly

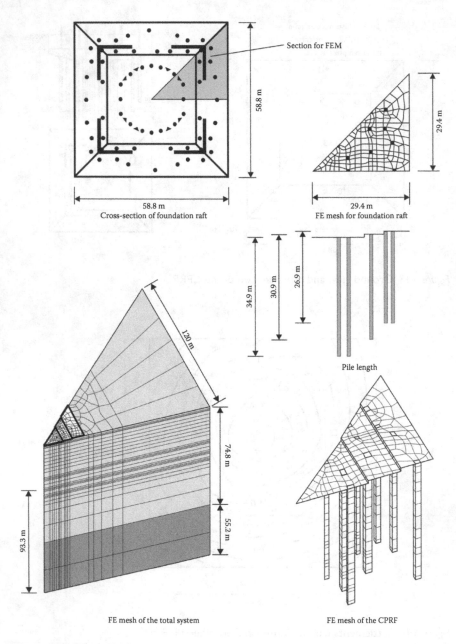

Figure 5.15 FE-mesh of the foundation.

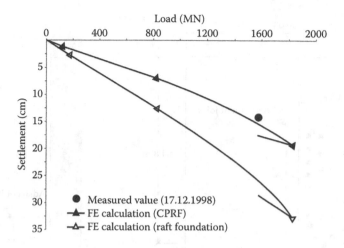

Figure 5.16 Measured and calculated settlements.

equal to the *in situ* measured maximum settlements of 13 cm. The CPRF coefficient is about $\alpha_{CPRF} = 0.43$ [24].

Based on several pile load tests in Frankfurt Clay, the ultimate skin friction of bored piles was estimated to 60–80 kN/m² for 20 m long piles. At the piles of the Messeturm, an average skin friction q_s of 90 kN/m² to 105 kN/m² was measured. At the pile toe, a maximum skin friction of $q_s = 200$ kN/m² was measured.

A pure pile foundation would have required 316 piles with 30 m length. In comparison to the realized CPRF with 64 piles and an average length of about 30 m, a pure pile foundation would have required more resources—such as, concrete and energy—and would have been about 3.9 million Euros more expensive than the CPRF.

5.6.3 DZ-Bank

The building complex of the DZ-Bank is situated in the financial district of Frankfurt am Main, Germany. It is a 208 m high office tower with 53 storeys and a 12-storey apartment building surrounding the skyscraper in an L-shape on two sides (Figure 5.17) [25]. Due to its high slenderness ratio (H/B = 4.7), the office tower with a total structural load of 1420 MN is founded on a CPRF in the Frankfurt Clay to reduce the risk of differential settlements [26]. The foundation consists of a basement with three underground levels, a raft with a thickness of 3.0–4.5 m and 40 bored piles with a diameter of 1.3 m and with a constant length of 30 m. The raft is founded at a depth of 14.5 m below the surface and is about 9.5 m below the groundwater level. The piles are concentrated beneath the heavy columns of the superstructure. The 2940 m² raft of the skyscraper is separated

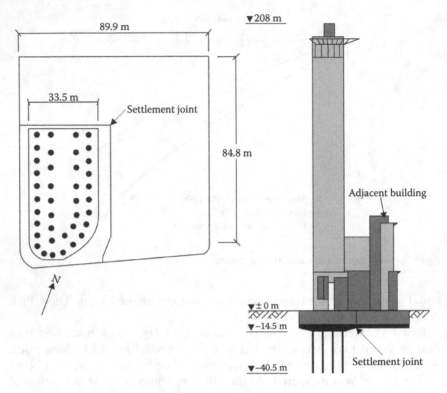

Figure 5.17 Ground view and cross-section of DG-Bank.

by a settlement joint from the adjacent raft of the neighboring building, which has a plan area of 3000 m² [27]. Hence the office tower is founded on its own centrically loaded CPRF.

The CPRF was monitored using geotechnical measuring devices and geodetic measurement points. Figure 5.18 shows the measured load-settlement curve for the foundation separated into the resistance of the piles and the raft. The increasing load of the building during the construction process (1990–1993) and the measured load distribution within the CPRF are illustrated in Figure 5.19, together with the measured settlements. The structural load is shared nearly equally by the raft and the piles, leading to a CPRF coefficient of $\alpha_{CPRF}=0.5$ [28]. This load sharing remained fairly constant during most of the construction period. At the completion of the concrete shell of the skyscraper, 9 cm of settlement was measured at the center of the raft. At the same time, the pile loads beneath the raft of 9.2–14.9 MN were measured, while beneath the raft, the effective contact pressure was measured at 150 kN/m² and the pore water pressure was 50 kN/m².

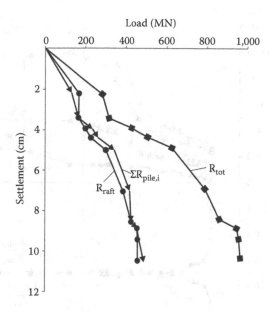

Figure 5.18 Measured load-settlement curves of CPRF.

5.6.4 American Express

The American Express office building in Frankfurt am Main, Germany, is 74 m high and was constructed in 1991–1992. The raft is loaded eccentrically by the 16-storey office tower (Figure 5.20). The building is founded on one single raft without any settlement joints between the office tower and the surrounding apartment buildings. To minimize tilting and differential settlements of the raft, 35 bored piles with a diameter of 0.9 m and a length of 20 m were located under the tower. Hence the foundation of the American Express building is the first example of a CPRF in Germany where the resistance of the foundation is centralized by concentrating piles under the eccentrically loaded area of the raft without any settlement joints in the raft [29].

5.6.5 Japan Center

The 115.3 m high Japan Center office tower is located in the center of the financial district in the west of Frankfurt am Main, Germany. The building comprises four basement floors and a tower with 29 floors and a ground view of 36.6 m×36.6 m (Figure 5.21). It is founded on a CPRF in a depth of 15.8 m below the surface. The total structural load is about 1050 MN. The raft has a thickness of 3.0 m at the center, reducing to 1.0 m at the edges. The raft is loaded with a remarkable eccentricity in the total building load of 7.5 m. Therefore, the positions of the 25 bored piles (diameter

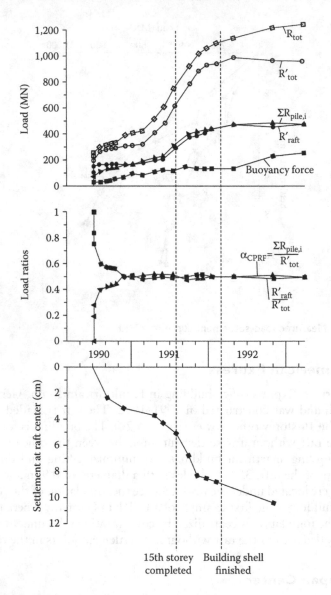

Figure 5.19 Observed time-dependent load-settlement behavior and load distribution of the CPRF.

of 1.3 m, length of 22 m) under the raft were optimized during the design to guarantee constant settlements over the entire foundation.

At the site of the Japan Center, the transition between the Frankfurt Clay and the rocky Frankfurt Limestone is located approximately 43 m below the surface, which is only about 5 m below the base level of the piles. Hence, the load-bearing behavior of the CPRF is influenced by the stiff

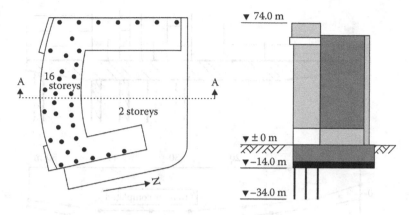

Figure 5.20 Ground view and cross-section A-A of American Express.

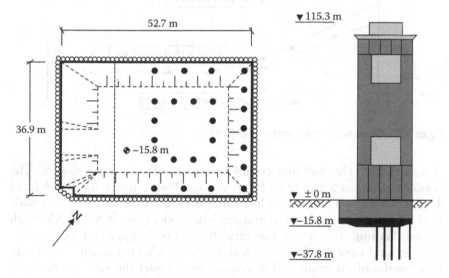

Figure 5.21 Ground view and cross-section of Japan Center.

limestone stratum. The CPRF was monitored using geotechnical measuring devices and geodetic measurement points. Figure 5.22 shows the measured settlements of the raft up to 6 months after the end of construction of the concrete shell. The basement settlements were 1.9–3.6 cm [30]. The foundation raft transfers about 60% of the structural load, the piles carry 40%. The measured pile loads are between 7.9–13.8 MN.

5.6.6 Kastor and Pollux

The building complex Kastor and Pollux (Forum Frankfurt) is located 150 m southeast of the Messeturm in comparable ground conditions

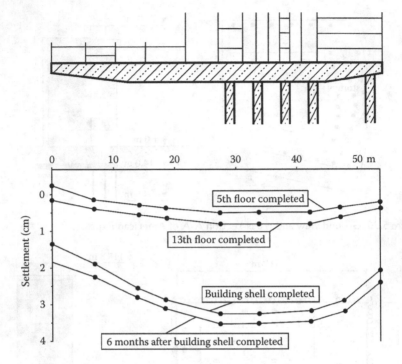

Figure 5.22 Observed settlements of the CPRF.

(Figure 5.23). The building complex consists of two office towers. The Kastor tower has a height of 94 m and the Pollux tower has a height of 130 m. These two towers are located at opposite ends of a 120.5 m wide parking basement with three underground floors (Figure 5.24). Although the raft loading is extremely eccentric, the raft is designed as a single structure (ground view of 14000 m²) with bored piles with a diameter of 1.3 m and lengths of 20 m and 30 m concentrated under the Kastor tower (26 piles) and under the Pollux tower (22 piles). The thickness of the foundation raft is about 3.0 m beneath the towers and 1.0 m in the area of the parking basement [30,31].

When the concrete shells of both towers were finished in 1996, the settlements in the parking area were between 4 cm and 6 cm, and under the towers the settlements were between 6 cm and 7 cm (Figure 5.25). The CPRF coefficient was $\alpha_{CPRF} = 0.35 - 0.4$.

5.6.7 Treptowers

The Treptowers in Berlin, Germany, is a 121 m high office building of the Allianz insurance company. It is located in the vicinity of the river Spree (Figure 5.26). The subsoil consists of fillings up to a depth of 3 m below the

Figure 5.23 Messeturm (left), Kastor (center) and Pollux (right).

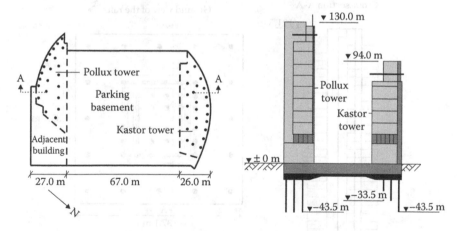

Figure 5.24 Ground view and cross-section A-A.

surface. Below the fillings, loose and medium dense sands follow down to a depth of 40 m. The lower surface of the basement is 8 m below the surface in the area of the elevator pit (Figure 5.27).

The high-rise building is founded on a CPRF with 54 bored piles with a diameter of 0.88 m. Depending on their location, the length of the reinforced-concrete piles varies from 12.5 m to 16 m. To improve the skin friction, a shaft grouting at the piles was performed. The raft has a size of 37.1 m × 37.1 m and a thickness of 2–3 m.

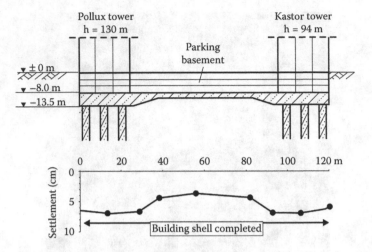

Figure 5.25 Measured settlement profile of CPRF.

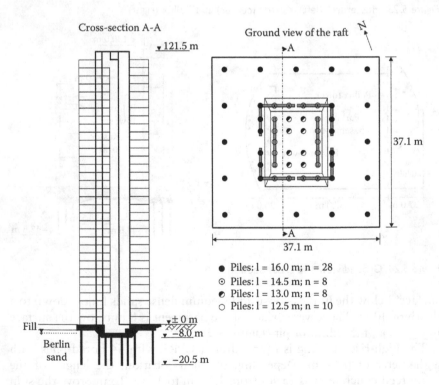

Figure 5.26 Treptowers.

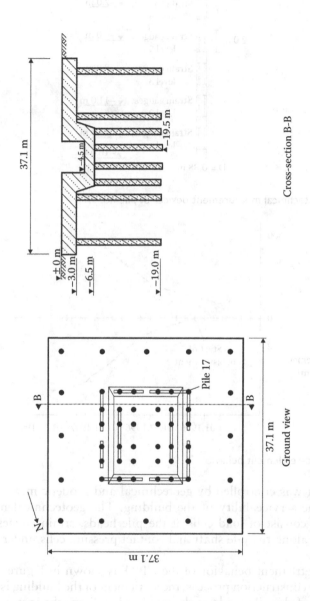

Cross-section B-B

Ground view

Figure 5.27 Ground view and cross-section B-B.

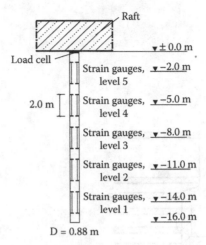

Figure 5.28 Geotechnical measurement devices of pile no. 17.

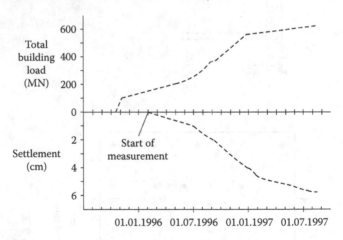

Figure 5.29 Load-settlement behavior.

The project was controlled by geotechnical and geodetic measurements to observe the serviceability of the building. The geotechnical measurement devices consist of load cells at the pile heads, strain gauges at different depths along the pile shaft and contact pressure cells under the raft (Figure 5.28).

The load-settlement behavior of the CPRF is shown in Figure 5.29. At the end of the construction process, the settlement of the building is 6.3 cm. The increase of the pile loads at the pile heads follows the increase of the building loads (Figure 5.30).

To analysis the load-settlement behavior numerical analysis have been carried out for the CPRF. The results of the numerical analysis were

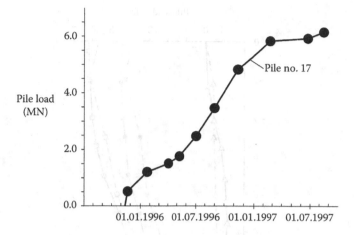

Figure 5.30 Measured pile load.

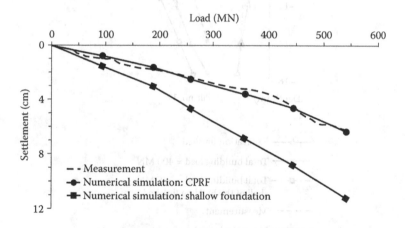

Figure 5.31 Load-settlement behavior of the CPRF.

compared with the results of the geotechnical and geodetic settlement measurements shown in Figures 5.31 and 5.32. In Figure 5.31, all loads and resistances are related to the beginning of the geodetic settlement measurements. At this stage of the construction process, the total building load had been 135 MN. The measured and calculated load-settlement behavior of the CPRF is in good accordance for the whole loading process. The CPRF coefficient is $\alpha_{CPRF} = 0.65$.

The numerical simulation for a spread foundation showed settlements of 11.1 cm (Figure 5.31). The CPRF reduced the settlements to 57% of the settlements of a spread foundation and led to a significant reduction of the differential settlements between the raft and the neighboring area [32].

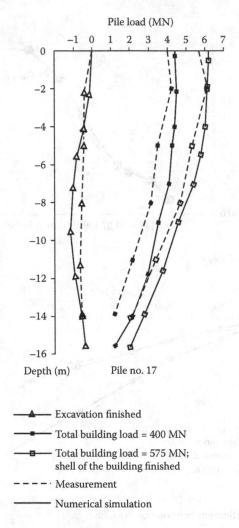

Figure 5.32 Load distributions along the pile.

In Figure 5.32, the load distribution along pile no. 17 is shown after the excavation of the pit and for total building loads of 400 MN and 575 MN after the end of construction of the shell of the building. All loads are related to the stage of pile installation. The negative pile loads results from the excavation and the reduced stress level. This effect has been observed at other projects as well [33]. For the 26.9 m to 34.9 m long piles, the negative pile loads have been estimated to 1 MN. Comparison of the calculation results and the measurements shows quite a good accordance.

5.6.8 Main Tower

The Main Tower is a high-rise building of 198 m with 5 basement levels and 57 levels above the ground in Frankfurt am Main, Germany (Figure 5.33). The total load of the building is 1900 MN. The raft, with its thickness of 3.0 m to 3.8 m, is founded 21 m below the surface and 14 m below the groundwater level. The Main Tower was constructed by the top-down method from 1996 to 1999. This means, that after the construction of the retaining system and the temporary and permanent deep foundation elements, the excavation and construction of the basements went parallel to the construction of the superstructure (Figure 5.34).

The building shaft is arranged asymmetrically on the raft. The design of the foundation was determined by the requirement to reduce the settlements of the tower itself and of the surrounding buildings in order to ensure

Figure 5.33 Main Tower.

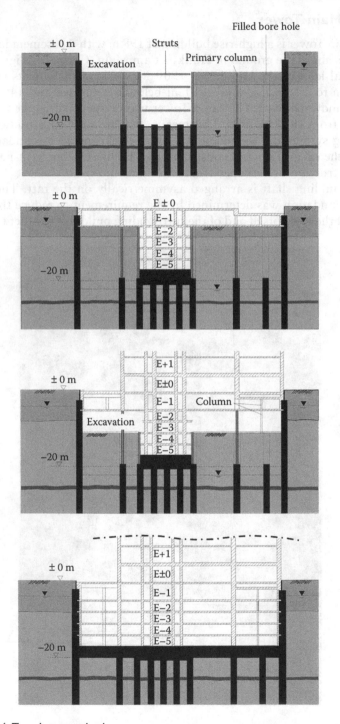

Figure 5.34 Top-down method.

their serviceability [34,35]. The Main Tower is founded on a CPRF with 112 bored piles with a diameter of 1.5 m and a length of 30.0 m. The pile load is transferred into the Frankfurt Clay, as the pile toes are located about 3 m to 8 m above the Frankfurt Limestone. The retaining system consisted of a secant pile wall of 257 bored piles with a diameter of 0.9 m, resp. 1.5 m. The project was controlled by geotechnical and geodetic measurements to observe the soil–structure interaction of the CPRF and the retaining system (Figure 5.35).

The horizontal deformation of the retaining walls was controlled by 14 inclinometers positioned behind the walls. The vertical deformation was measured down to a depth of 140 m below ground surface with 17 extensometers. Aim of the measurements carried out in the foundation piles was the monitoring of their bearing behavior as a pile foundation during the top-down method and later on as part of the CPRF. For this purpose, 17 piles were equipped with load cells at the pile toe and 14 piles with load cells at the pile head. To determine the load distribution along the pile shaft, 335 strain gauges were installed in 21 piles.

5.6.9 Sony Center

The 103 m high Sony Center office tower is part of a building complex at the Potsdamer Platz in Berlin, Germany. The tower, with a ground view of 2600 m², was constructed from 1998 to 2000 and is founded on a CPRF in difficult soil conditions. The soil investigations showed a layer of semi-solid glacial drift, which is located at a depth of 11–12 m below the surface, just beneath the raft [36,37]. The CPRF consists of a 1.5–2.5 m thick raft and 44 bored piles with a diameter of 1.5 m and lengths of 20–25 m (Figure 5.36).

5.6.10 Victoria-Turm

CPRF offers a technical and economical optimal solution not only for complex constructions with a sensible neighboring construction or difficult ground conditions. The Victoria-Turm in Mannheim, Germany, with a height of 97 m, was founded on a CPRF (Figure 5.37). The foundation raft has a thickness of 3 m. The length of the piles is 20 m in the center and 15 m at the edges.

5.6.11 City Tower

In the densely populated inner city of Offenbach am Main, Germany, the 140 m high City Tower was founded on a CPRF in settlement-sensitive clay [38,39]. At a distance of only 4 m, a railway tunnel is situated parallel to the project area (Figure 5.38). For the calculation of the CPRF, numerical simulations were done, taking into account the symmetry axis. The load

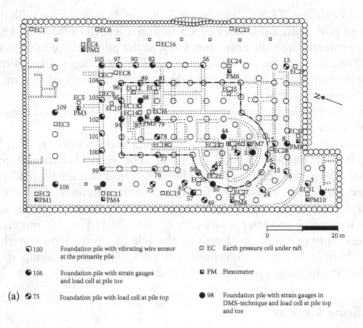

(a)

● 100	Foundation pile with vibrating wire sensor at the primarily pile	□ EC	Earth pressure cell under raft
● 106	Foundation pile with strain gauges and load cell at pile toe	◙ PM	Piezometer
◐ 75	Foundation pile with load cell at pile top	● 98	Foundation pile with strain gauges in DMS-technique and load cell at pile top and toe

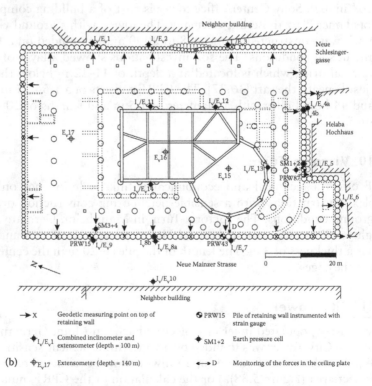

(b)

→ x	Geodetic measuring point on top of retaining wall	◗ PRW15	Pile of retaining wall instrumented with strain gauge
✦ I_v/E_v1	Combined inclinometer and extensometer (depth = 100 m)	✦ SM1+2	Earth pressure cell
⊕ E_v17	Extensometer (depth = 140 m)	←→ D	Monitoring of the forces in the ceiling plate

Figure 5.35 Instrumentation of CPRF (a) and the retaining system (b).

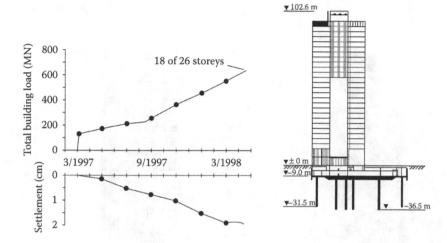

Figure 5.36 Sony Center.

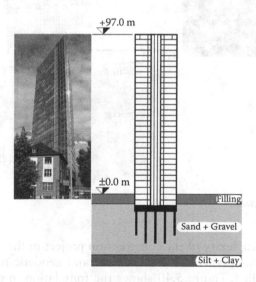

Figure 5.37 Victoria-Turm.

history of the project area in particular was investigated by simulating the deconstruction of the existing building, the excavation for the new construction, and the construction of the basement and the superstructure. Figure 5.39 shows the FE model of the optimized CPRF.

The CPRF consists of 36 piles with a length between 25 m at the edges and 35 m in the center. The diameter of the piles is 1.5 m. The foundation raft has a thickness of about 3 m.

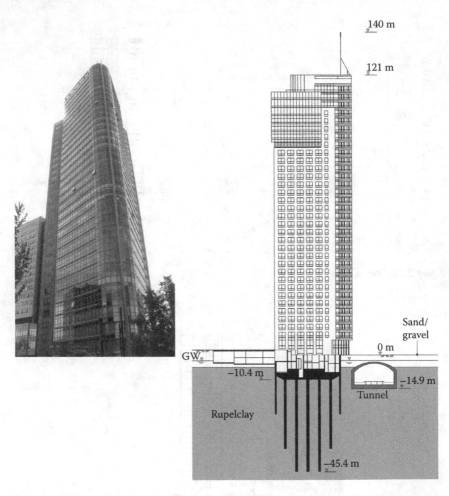

Figure 5.38 City Tower.

Due to the complexity of the construction project in the area of the railway tunnel, a comprehensive geotechnical and geodetic monitoring program was installed. Figure 5.40 shows the foundation in ground view as well as the associated geotechnical instrumentation. Six foundation piles were instrumented with load cells at the piled head and the pile toe as well as eight strain gauges in four different depths along the pile. The generated settlements from the new building were measured by extensometers under the core of the high-rise building and an extensometer between the retaining system and the railway tunnel. Moreover, an inclinometer was installed to measure the horizontal deformations behind the retaining system in the area of the tunnel. Figure 5.41 illustrates schematically the application of load cells, soil pressure sensors and piezometers.

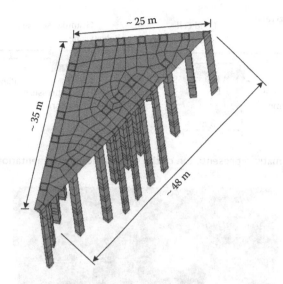

Figure 5.39 FE-mesh of the CPRF using symmetry axis.

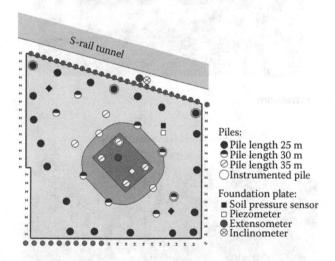

Piles:
● Pile length 25 m
◓ Pile length 30 m
⊘ Pile length 35 m
○ Instrumented pile

Foundation plate:
■ Soil pressure sensor
□ Piezometer
● Extensometer
⊗ Inclinometer

Figure 5.40 Ground view of CPRF and the geotechnical instrumentation.

5.6.12 Darmstadtium

The result of the soil investigation for the science and congress center Darmstadtium in Darmstadt, Germany, showed that the planned
construction is situated above the eastern fault of the Rhine Valley. The
Darmstadtium was opened in 2007. The finalized project is shown in
Figure 5.42. The Rhine Valley fault crosses the project area as shown
in Figure 5.43 [40]. In the northern and western area, unconsolidated

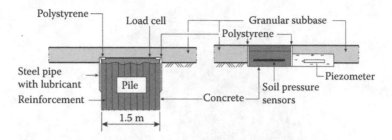

Figure 5.41 Schematic representation of the geotechnical instrumentation.

Figure 5.42 Darmstadtium.

Figure 5.43 Excavation pit.

sediments of the Rhine Valley fault were found. In the eastern and southern area, rocks of the Odenwaldcrystalline were identified (granodiorite). Up to now, tectonic activities along the fault zone have not finished. The areas of Darmstadt that are located west of the Rhine Valley fault settle about 0.5 mm per year. The foundation system and the rising structure have to

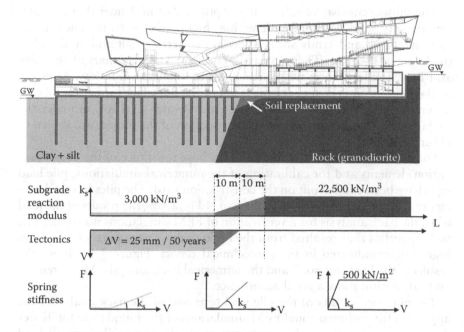

Figure 5.44 Foundation system.

be designed for these tectonic displacements. The foundation in the area of the rock was constructed as a spread foundation. In the area of the Rhine Valley fault, the foundation was constructed as a CPRF (Figure 5.44).

5.6.13 Mirax Plaza

The Mirax Plaza in Kiev, Ukraine, consists of two high-rise buildings, each of them with a height of 192 m (46 storeys), as shown in Figure 5.45. The ground view of the project is about 294000 m².

Figure 5.45 Mirax-Plaza.

The subsoil consists of filling to a depth of 2–3 m. Under that are quaternary silty sand and sandy silt with a thickness of 5–10 m. The quaternary silty sand and sandy silt is underlain by tertiary silt and sand with a thickness of 0–24 m. Then tertiary, clayey silt and clay marl of the Kiev and Butschak formation with a thickness of about 20 m follow. After the tertiary clayey silt, the tertiary fine sand of the Butschak formation follows down to the investigation depth. The groundwater level is 2 m under the surface. The soil conditions and a cross-section of the project are shown in Figure 5.46.

For the verification of the shaft and the base resistance of the deep foundation elements and for calibration of the numerical simulations, pile load tests have been carried out on the construction yard. The piles have a diameter of 0.82 m and a length of 10 m, resp. 44 m. Using the results of the load tests, the back analysis for a verification of FEM simulations was done. The soil properties that resulted from the back analysis were partly three times higher than indicated in the geotechnical report. Figure 5.47 shows the results of the load test no. 2 and the numerical back analysis. Measurement and calculation show a good accordance.

The obtained results of the pile load tests and of the back analysis were applied in three-dimensional FEM simulations of the foundation for Tower A, taking advantage of the symmetry of the building. The overall load of Tower A is about 2200 MN, and the area of the foundation is about 2000 m² (Figure 5.48).

The foundation design considers a CPRF with 64 barrettes with 33 m length and a cross-section of 2.8 m × 0.8 m. The raft of 3 m thickness is located in Kiev clay marl at about 10 m depth below the surface. The barrettes are penetrating the layer of Kiev clay marl reaching the Butschak sands.

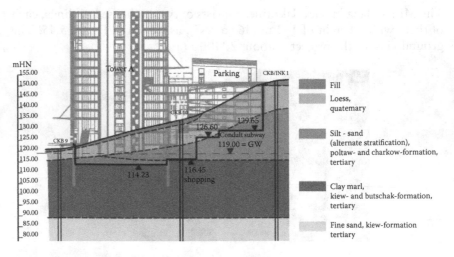

Figure 5.46 Soil conditions and cross-section of the project area.

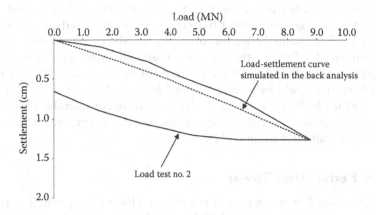

Figure 5.47 Results of the *in situ* load test and the numerical simulations.

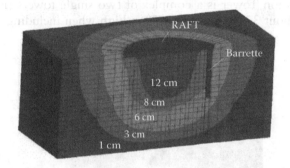

Figure 5.48 FE-model of CPRF of Tower A and calculated settlements in centimeters.

The calculated loads on the barrettes were in the range of 22.1–44.5 MN. The load on the outer barrettes was about 41.2–44.5 MN, which significantly exceeds the loads on the inner barrettes with the maximum value of 30.7 MN. This behavior is typical for a CPRF. The deep foundation elements, which are located at the edge of the foundation raft, get more of the total load because of their higher stiffness due to the higher volume of the activated soil. The CPRF coefficient is $\alpha_{CPRF}=0.88$. Maximum settlements of about 12 cm were calculated due to the settlement-relevant load of 85% of the total design load. The pressure under the foundation raft is calculated in the most areas as not exceeding 200 kN/m². At the raft edge, the pressure reaches 400 kN/m². The calculated base pressure of the outer barrettes has an average of 5100 kN/m² and for inner barrettes an average of 4130 kN/m². The mobilized skin friction increases with the depth reaching 180 kN/m² for outer barrettes and 150 kN/m² for inner barrettes.

During the construction of Mirax Plaza, the observational method according to EC 7 was applied. In particular, the distribution of the loads between the barrettes and the raft was monitored. For this reason, three earth pressure devices were installed under the raft and two barrettes (most loaded outer barrette and average loaded inner barrette) were instrumented over the length.

The CPRF of the high-rise building project Mirax Plaza represents the first authorized CPRF in the Ukraine. Using the advanced optimization approaches and taking advantage of the positive effect of CPRF, the number of barrettes could be reduced from 120 with 40 m length to 64 with 33 m length. The foundation optimization leads to considerable decrease of the utilized resources (cement, aggregates, water, energy, etc.) and to cost savings of about €2.4 million [41].

5.6.14 Federation Tower

The Federation Tower is a part of the project Moscow City, which contains the construction of several high-rising buildings for business in Moscow, Russia, on an area of more than one square kilometer [42].

The Federation Tower is a complex of two single towers (Figure 5.49). Tower A is about 374 m high, or 450 m high when including the spire on

Figure 5.49 Federation Tower.

the roof. Tower B is about 243 m high. Construction started in 2003. Due to the geometry, the high-rise double-towers surpassed at that time all previous experiences with high-rise buildings in Russia and in Europe. The two towers will share a 4.6 m thick foundation raft about 140 m long and 80 m wide, made of reinforced concrete at a depth of about 20 m below the surface.

The total load is about 3000 MN for Tower A and 2000 MN for Tower B. With additional loads of about 1000 MN for adjacent buildings and the basement floors, and a load of about 1300 MN for the foundation raft itself, the total load is about 7300 MN.

The project area of Moscow City is located to the west of the central district of Moscow on the left bank of the River Moskva. Underneath anthropogenically influenced artificial backfill, at first the quaternary accumulation of the river terrace and underneath the alternating sequences of the carbon are located. Underneath the foundation level of the Federation Tower, a complex alternating sequence is found, consisting of variably intensively fissured, cavernous and porous limestone and variably hard, more or less watertight clay (marl). The layers have different thicknesses with a range of 3–10 m. Moscow City and also the construction area of the Federation Tower are located in a territory, where potentially dangerous karst-suffusion processes occur.

In the project area, several groundwater horizons carrying confined water, have been found. Because of the sealing effect of the clay (marl), they are not or just moderately corresponding with each other. The groundwater mainly circulates in the fissured and karst-suffusion affected limestone. The groundwater, circulating in the lower limestone-aquifer, has a water pressure of about 12 m. The limestone horizons located above, may have groundwater with higher hydraulic pressures.

Two pile load tests TP-15-1 and TP-15-2 using O-cells have been carried out in 2005. The test piles were constructed with a diameter of 1.2 m and a length of 6.90 m and 13.35 m. The empty drill hole was filled with sand. The piles are completely positioned in the limestone (Figure 5.50).

Each pile was divided into two segments. Between the pile segments O-cells were installed. In each test pile, displacement transducers were installed to measure the displacements of the pile segments.

Pile load test TP-15-1 was carried out to a maximum load of about 33 MN. In between, an unloading-phase was made at 15 MN back to zero-load, followed by a reloading-phase (Figure 5.51). The final displacement was about 0.6 cm at the upper pile segment and 0.4 cm at the lower pile segment. Neither was a failure of one of the pile segments seen, nor the empirically defined limit of $0.1*D = 12$ cm reached. The evaluation of the pile load test TP-15-1 yields a skin friction of 1140 kN/m^2 and a base resistance of 5380 kN/m^2. These values are not ultimate ones.

Pile load test TP-15-2 was carried out to a maximum load of about 33 MN as well. In between, three unloading-phases back to zero-load were

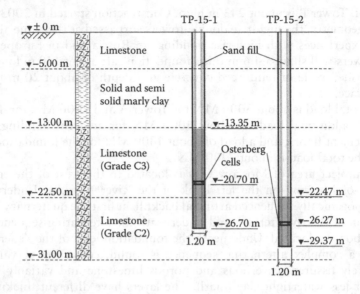

Figure 5.50 Section of test piles TP-15-1 and TP-15-2.

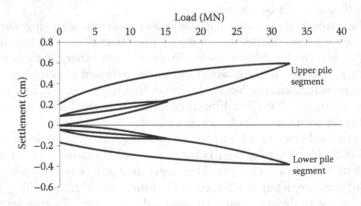

Figure 5.51 Load-displacement diagram of test pile TP-15-1.

made (Figure 5.52). The final displacement was about 4.3 cm at the upper pile segment and 2.2 cm at the lower pile segment. Again, neither was a failure of one of the pile segments seen, nor the empirically defined limit of $0.1*D = 12$ cm reached. The evaluation of the pile load test TP-15-2 yields a skin friction of 2310 kN/m² and a base resistance of about 5630 kN/m².

5.6.15 Exhibition Hall 3

Exhibition Hall 3 in Frankfurt am Main, Germany, was finished in 2001 and is one of the biggest exhibition halls in Europe. The hall has a length of

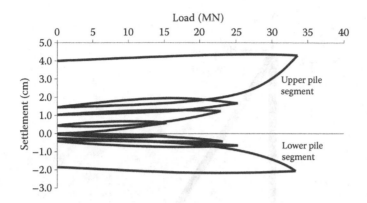

Figure 5.52 Load-displacement diagram of test pile TP-15-2.

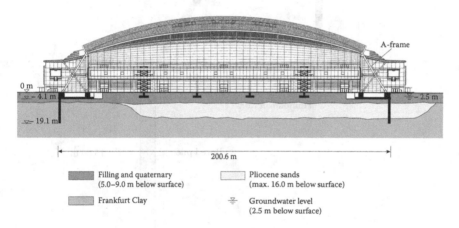

Figure 5.53 Cross-section and subsoil conditions.

about 210 m, a width of about 130 m, and a height of about 45 m. The roof, with a free span of 165 m, was designed as a double-curved, three-dimensional, load-bearing structure consisting of five arched compression trusses and six arched tension trusses. A cross-section of the realized project is shown in Figure 5.53. Twelve A-frames, six at each side, carry the horizontal and the vertical loading of the roof (Figure 5.54). These A-frames, with a height of 24 m, are constructed of two vertical steel tubes [43,44].

The subsoil and groundwater conditions are visualized in Figure 5.53. Beneath the surface, a layer of fill and quaternary soil with a thickness between 5 m and 9 m exists. This layer is followed by tertiary sediments. During the soil investigation, a soil layer of tertiary sand and gravel was detected. It crosses the project area in diagonal direction. Under this tertiary sand and gravel follows the Frankfurt Clay.

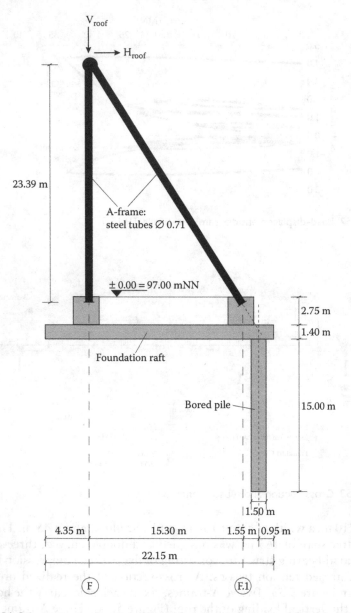

Figure 5.54 Schematic visualization of the A-frame and the horizontal-loaded CPRF.

Due to the strong interaction between the superstructure, the foundation, and the subsoil, a strong limitation of the displacements of the foundation is necessary. The design of the horizontal-loaded CPRF is based on three-dimensional numerical analysis. Each of the two CPRFs (one on each end of the hall) consists of a raft and 14 bored piles. The raft has a thickness of 1.4 m, a length of 127.5 m, and a width of 22.15 m. The bored piles have a diameter of 1.5 m and a length of 15 m.

The horizontal displacements of the CPRF were observed with four inclinometers reaching 50 m under the surface. To measure the vertical displacements, four extensometers were installed. To complete the necessary monitoring program, geodetic measurement points were defined and pressure cells in the soil and strain gauges at the A-frames were placed. The measured horizontal displacements are about 1 cm. The measured vertical displacements are about 1–3.5 cm.

The example shows that the CPRF can be used for a settlement-reduced transfer of horizontal loads into the subsoil.

REFERENCES

1. Deutsche Gesellschaft für Geotechnik e.V. (2012): *Empfehlungen des Arbeitskreises Pfähle(EA-Pfähle)*. 2. Auflage, Ernst & Sohn Verlag, Berlin.
2. Hanisch, J.; Katzenbach, R.; König, G. (2002): *Kombinierte Pfahl-Plattengründungen*. Ernst & Sohn Verlag, Berlin.
3. International Society for Soil Mechanics and Geotechnical Engineering (2013): ISSMGE Combined Pile-Raft Foundation Guideline, Darmstadt, Germany.
4. Katzenbach, R.; Boled-Mekasha, G.; Wachter, S. (2006): *Gründung turmartiger Bauwerke. Beton-Kalender*. Ernst & Sohn Verlag, Berlin, 409–468.
5. Amann, P. (1975): Über den Einfluss des Verformungsverhaltens des Frankfurter Tons auf die Tiefenwirkung eines Hochhauses und die Form der Setzungsmulde. Mitteilungen der Versuchsanstalt für Bodenmechanik und Grundbau der Technischen Hochschule Darmstadt, Germany, Heft 23.
6. Cooke, R.W. (1986): Piled raft foundations on stiff clays: A contribution to design philosophy. *Géotechnique* 36, No. 22, 169–203.
7. Poulos, H.G. (1989): Pile behaviour: Theory and application. *Géotechnique* 39, No. 3, 365–415.
8. Randolph, M.F. (1994): Design methods for pile groups and piled rafts. *XIII International Conference on Soil Mechanics and Foundation Engineering*, 5–10 January, New Delhi, Vol. 5, 61–82.
9. El-Mossallamy, Y. (1996): Ein Berechnungsmodell zum Tragverhalten der Kombinierten Pfahl-Plattengründung. Mitteilungen des Institutes und der Versuchsanstalt für Geotechnik der Technischen Hochschule Darmstadt, Germany, Heft 36.
10. Poulos, H.G.; Small, J.C.; Ta, L.D.; Simha, J.; Chen, L. (1997): Comparison of some methods for analysis of piled rafts. *14th International Conference on Soil Mechanics and Geotechnical Engineering*, 6–12 September, Hamburg, Germany, Vol. 2, 1119–1124.
11. Katzenbach, R.; Reul, O. (1997): Design and performance of piled rafts. *XIV International Conference on Soil Mechanics and Foundation Engineering*, 6–12 September, Hamburg, Germany, Vol. 4, 2253–2256.
12. Horikoshi, K.; Randolph, M.F. (1998): A contribution to optimal design of piled rafts. *Géotechnique* 48, London, No. 3, 301–317.

13. Russo, G.; Viggiani, C. (1998): Factors controlling soil–structure interaction for piled rafts. *Darmstadt Geotechnics* No. 4, Darmstadt, Germany, Vol. 2, 297–321.
14. Deutsches Institut für Normung e.V. (2010): DIN 1054 Subsoil: Verification of the safety of earthworks and foundations—Supplementary rules to DIN EN 1997-1. Beuth Verlag, Berlin.
15. Deutsches Institut für Normung e.V. (2006): DIN 4017 Soil: Calculation of design bearing capacity of soil beneath shallow foundations. Beuth Verlag, Berlin.
16. Katzenbach, R.; Leppla, S.; Ramm, H.; Seip, M.; Kuttig, H. (2013): Design and construction of deep foundation systems and retaining structures in urban areas in difficult soil and ground water conditions. *11th International Conference, Modern Building Materials, Structures and Techniques*, 16–17 Mai, Vilnius, Lithuania, 540–548.
17. Reul, O. (2000): In-situ-Messungen und numerische Studien zum Tragverhalten der Kombinierten Pfahl-Plattengründung. Mitteilungen des Institutes und der Versuchsanstalt für Geotechnik der Technischen Universität Darmstadt, Germany, Heft 53.
18. Katzenbach, R.; Leppla, S.; Vogler, M.; Kuttig, H.; Dunaevskiy, R. (2009): Foundation optimization of high-rise buildings in Kiev. *Darmstadt Geotechnics*, No. 17, 81–96.
19. Turek, J. (2006): Beitrag zur Klärung des Trag- und Verformungsverhaltens horizontal belasteter Kombinierter Pfahl-Plattengründungen. Mitteilungen des Institutes und der Versuchsanstalt für Geotechnik der Technischen Universität Darmstadt, Germany Heft 72.
20. Sommer, H.; Wittmann, P.; Ripper, P. (1985): Piled raft foundation of a tall building in Frankfurt clay. *11th Conference of Soil Mechanics and Foundation Engineering*, San Francisco, CA, 2253–2257.
21. Sommer, H. (1986): Kombinierte Pfahl-Plattengründung eines Hochhauses in Ton. Vorträge der Baugrundtagung in Nürnberg, Nürnberg, Germany, 391–405.
22. Breth, H. (1970): Die Tragfähigkeit von Bohrpfählen im Frankfurter Ton. Das Tragverhalten des Frankfurter Tons bei im Tiefbau auftretenden Beanspruchungen. Mitteilungen der Versuchsanstalt für Bodenmechanik und Grundbau, Technische Hochschule Darmstadt, Germany, Heft 4, 51–69.
23. Reul, O. (2000): In-situ-Messungen und numerische Studien zum Tragverhalten der Kombinierten Pfahl-Plattengründung. Mitteilungen des Institutes und der Versuchsanstalt für Geotechnik der Technischen Universität Darmstadt, Germany, Heft 53, 136–154.
24. Sommer, H.; Katzenbach, R.; DeBenedittis, C. (1990): Last-Verformungsverhalten des Messeturms Frankfurt/Main. Vorträge der Baugrundtagung in Karlsruhe, Karlsruhe, Germany, 371–380.
25. Lutz, B.; Wittmann, P.; Theile, V. (1993): Modellierung von Gründungen im Hochhausbau am Beispiel ausgewählter Frankfurter Hochhäuser. Baustatik/ Baupraxis, Technische Universität München, 1–18 February, Germany.
26. Franke, E.; Lutz, B.; El-Mossallamy, Y. (1994): Measurements and numerical modelling of high rise building foundations on Frankfurt Clay. Conference on Vertical and Horizontal Deformations of Foundations and Embankments, ASCE Geotechnical Special Publication, Texas, No. 40, Vol. 2, 1325–1336.

27. Wittmann, P.; Ripper, P. (1990): Unterschiedliche Konzepte für die Gründung und Baugrube von zwei Hochhäusern in der Frankfurter Innenstadt. Vorträge der Baugrundtagung in Karlsruhe, Germany, 381–397.
28. Franke, E.; Lutz, B.; El-Mossallamy, Y. (1994): Pfahlgründungen und Interaktion Bauwerk-Baugrund. *Geotechnik*, 157–172.
29. Katzenbach, R.; Arslan, U.; Moormann, C. (2000): *Piled Raft Foundation Projects in Germany. Design Applications of Raft Foundations*, MPG Books. Bodmin, Cornwall, Great Britain, 323–392.
30. Lutz, B.; Wittmann, P.; El-Mossallamy, Y.; Katzenbach, R. (1996): Die Anwendung von Pfahl-Plattengründungen: Entwurfspraxis, Dimensionierung und Erfahrungen mit Gründungen in überkonsolidierten Tonen auf der Grundlage von Messungen. Vorträge der Baugrundtagung in Berlin, 153–164.
31. Katzenbach, R.; Hoffmann, H.; Vogler, M.; Moormann, C. (2001): Costoptimized foundation systems of high-rise structures, based on the results of actual geotechnical research. *International Conference on Trends in Tall Buildings*, Frankfurt, Germany, ed. König, Graubner, 421–443.
32. Richter, T.; Reul, O.; Arslan, U. (1998): Setzungen hoch belasteter Gründungen in Berliner Böden: Vergleich von Tief- und Flachgründungen in Berechnung und Messung. Vorträge der Baugrundtagung in Stuttgart, Germany, 1–18.
33. Sommer, H. (1993): Development of locked stresses and negative shaft resistance at the piled raft foundation: Messeturm Frankfurt am Main. *Deep Foundations on Bored and Auger Piles*, Rotterdam, the Netherlands, 347–349.
34. Katzenbach, R.; Schmitt, A.; Turek, J. (1999). Co-operation between the geotechnical and structural engineers: Experience from projects in Frankfurt. *Conference COST Action C7, Soil–Structure Interaction in Urban Civil Engineering*, 1.2, Thessaloniki, Greece.
35. Moormann, C. (2002) Trag- und Verformungsverhalten tiefer Baugruben in bindigen Böden unter besonderer Berücksichtigung der Baugrund-Tragwerk- und der Baugrund-Grundwasser-Interaktion. Mitteilungen des Institutes und der Versuchsanstalt für Geotechnik der Technischen Universität Darmstadt, Germany, Heft 59.
36. Richter, T.; Savidis, S.; Katzenbach, R.; Quick, H. (1996): Wirtschaftlich optimierte Hochhausgründungen im Berliner Sand. Vorträge der Baugrundtagung in Berlin, 129–146.
37. Richter, T.; Reul, O.; Arslan, U. (1998): Setzungen hoch belasteter Gründungen in Berlinder Böden: Vergleich von Tief- und Flachgründungen in Berechnung und Messungen. Vorträge der Baugrundtagung in Stuttgart, Germany, 601–613.
38. Katzenbach, R.; Schmitt, A.; Turek, J. (2001): Setzungsarme Hochhausgründung neben einem Tunnel. Beratende Ingenieure, Krammer Verlag Düsseldorf AG, Berlin, Heft 7/8, 32–35.
39. Katzenbach, R.; Bachmann, G.; Gutberlet, C.; Schmitt, A.; Turek, J. (2003): Deep foundations: Combined pile-raft foundations of Frankfurt high-rise buildings. 5th Suklje Day, 9–11 June, Rogastka Slatina, Slovenia, 1–20.
40. Katzenbach, R.; Leppla, S.; Ramm, H.; Waberseck, T.; Vogler, M.; Seip, M. (2012): Geotechnik und Geothermie in der Region Rhein-Main-Neckar. 32. Baugrundtagung der Deutschen Gesellschaft für Geotechnik e.V., Mainz, Germany, 7–14.

41. Katzenbach, R.; Leppla, S. (2014): Deep foundation systems for high-rise buildings in difficult soil conditions. *Geotechnical Engineering Journal of the SEAGS & AGSSEA*, June, Vol. 45, No. 2, 115–123.
42. Katzenbach, R.; Leppla, S.; Vogler, M.; Dunaevskiy, R.; Kuttig, H. (2010): State of practice for the cost-optimized foundation of high-rise buildings. *International Conference Geotechnical Challenges in Megacities*, 7–10. June, Moscow, Russia, Vol. 1, 120–129.
43. Turek, J.; Katzenbach, R. (2004): New exhibition hall 3 in Frankfurt: Case history of a combined pile-raft foundation subjected to horizontal load. *5th International Conference on Case Histories in Geotechnical Engineering*, 13–17. April, New York, 1–5.
44. Katzenbach, R.; Leppla, S.; Ramm, H. (2014): Combined pile-raft foundation: Theory and practice. *Design and Analysis of Pile Foundations*, ed. Ali Bouafia, Dar Khettab Press, Boudouaou, Algeria, 262–291.

Chapter 6

Dynamic behavior of foundation systems

6.1 INTRODUCTION TO DYNAMIC ASPECT OF DEEP FOUNDATION SYSTEM

Deep foundations, like conventional pile foundations and the relatively new combined pile-raft foundation (CPRF), which is a hybrid of shallow and deep foundation, are widely used for both offshore and onshore structures to transfer superstructure load to the deeper strata. They are generally preferred where topsoil is loose, soft, and susceptible to shrinkage and swelling, when shallow foundations experience huge uplift pressure due to a fluctuating water table and are not safe under serviceability conditions. Though reasonably good progress was made for the design of pile foundations and CPRF under lateral loads for static loading, their performance under seismic loading is an area of concern to geotechnical practitioners over the last few decades due to increasing demand for high-rise structures. The problem is more complex if the soil is susceptible to liquefaction during earthquake. More attention needs to be given to pile foundations as floating piles passing through liquefiable soil layers undergo significant loss of shaft resistance. In the case of end-bearing piles, excessive loads may be transferred to the end-bearing strata due to loss of shaft resistance, and thus the pile is subjected to higher bearing pressures. Also, loss of shear strength in the liquefied zone will increase the effective length of the pile, and thus the pile may fail by buckling if axial loads are predominant. In the case of CPRF, the presence of liquefied soil may alter load-sharing aspects on different components of the foundation system.

Several examples of failure of pile foundations during earthquakes are available in the literature, for example, Figure 6.1: the failure of Showa Bridge and the Niigata Family Court House during the 1964 Niigata earthquake, the Landing Bridge in New Zealand during the 1897 Edgecumbe earthquake, the Harbour Master's Tower at Kandla Port in India during the 2001 Bhuj earthquake, and many more. The excellent performance of the Twelve storey building (Figure 6.2) and Hardon experimental hall in the 2011 Tohoku earthquake ($M_w = 9.0$) in Japan, which were founded on

Figure 6.1 Failure of Showa Bridge during Niigata 1964 earthquake: Lateral spreading cause bridge spans to fall. (From Kramer, S.L., Geotechnical earthquake engineering. In: Prentice-Hall international series in civil engineering and engineering mechanics. Prentice-Hall, New Jersey, 1996.)

Figure 6.2 Twelve storey building survived after Tohuku Earthquake (M_w=9.0) founded on CPRF. (From Yamashita et al., *Soils and Foundations*, 52, 1000–1015, 2012.)

CPRF, provided an insight for geotechnical practitioners to decide on the applicability of such foundation systems for high-rise buildings.

In order to evaluate the response of deep foundations, it is very important to understand the loss of bearing and lateral restrain in loose, saturated, non-cohesive soil and soft clay. The reduction in soil stiffness is due to loss of shear strength and the area of the liquefying zone needs to be evaluated properly. The response of deep foundations to earthquakes may lead to liquefaction, which involves complex material behavior such as increased pore water pressure, reduced effective strength, and stiffness degradation, as well as complex mass behavior of the ground such as kinematic loading and changes in the performance of the superstructure due to inertial loading. Advances in the field of computer engineering are helping geotechnical practitioners to simulate actual dynamic conditions and to record their reactions in a realistic way.

Two types of earthquake are envisioned in the seismic design of buildings: a medium earthquake, with an intensity of about V and recurrence interval of 50 years; and the largest possible earthquake, with an intensity of about VI to VII and recurrence interval of 500 years. Table 6.1 is the performance matrix used for the building design. The minimum legal requirement for a medium earthquake is that the building must be operational after the earthquake, and the minimum requirement for a building designed for the largest possible earthquake is that building should not collapse and or endanger the lives of its occupants.

Two design methods are used for the design of deep foundations for a building as per the height of the building. One is the requirement of the foundation structure alone and the other is the requirement for the behavior of the foundation structure to suppress the response of the building within the design requirements. Seismically induced loads are replaced by equivalent static horizontal load, and the load from the foundation is obtained from the weight of the foundation multiplied by the seismic coefficient. The horizontal load from the building is calculated, taking into consideration the natural frequency of the building. While calculating the induced stresses of the structural members of the foundation, the foundation structure alone is considered, regardless of the superstructure. Table 6.2 shows the basic requirements in the design of the building.

Table 6.1 Performance matrix of earthquake

Earthquake level (reoccurrence interval)	Minimum requirement of the performance level	
	Operational	Life safe
Medium (50 years)	Required	Required
Largest (500 years)	Not required	Required

Table 6.2 Basic requirements for the design of the building

Building height	Design recommendation
Height > 60 m	Dynamic analysis done for medium and largest possible earthquake
Height < 60 m	Allowable stress design (equivalent static external load applied to building)
31 m < height < 60 m	Load carrying capacity and deformation response should be within allowable value
Height < 30 m	Plastic deformation of the building permitted (building should not collapse)

This chapter focuses on the selection of dynamic soil parameters, which determine the dynamic response of soil in the design of deep foundations. It describes free-field ground response analysis, which shows the local site effect on amplification or de-amplification of the subsoil under different seismic loadings. It provides brief details about liquefaction and cyclic softening caused by strong shaking, and it provides guidelines to evaluate safety against liquefaction in case of loose, saturated, non-cohesive soil; it also provides liquefaction hazard mapping. It incorporates brief discussion about the design of pile foundations under pseudo-static and dynamic conditions. It also provides the various analyses adopted for the design of CPRF under seismic conditions and discusses various case studies for these foundations. It focuses on the seismic analysis of the well foundation for bridges. Finally, it discusses the building code provisions and some case studies.

6.2 DYNAMIC SOIL PARAMETERS

Damage due to an earthquake event is influenced by the response of soil to cyclic loading, which is dependent upon the engineering properties of the soil during the event. The evaluation of dynamic soil property is very important to study the site response of *in situ* soil and dynamic analysis of geotechnical structures. The behavior of soil under static and dynamic conditions is the most important factor in seismic foundation design. Geotechnical behavior criteria include plastic deformation characteristics, strength, damping, and stiffness of soil. Data on plastic deformation characteristics are needed to understand the earthquake-induced permanent displacement, strength characteristics are required for foundation stability analysis, and damping and stiffness are the response characteristics of the soil deposits. Duration and frequency content for seismic event influences all the characteristics of soil deposits and must be considered in the design. An accurate estimation of dynamic soil parameters provides confidence in estimating the local soil amplification during an earthquake event.

6.2.1 Determination of dynamic soil parameters

The determination of dynamic soil properties is critical, and many field and laboratory techniques are available to address the problem. Soil parameter that influences wave propagation includes stiffness, damping, Poisson's ratio, and density. Of these properties, wave propagation primarily depends on soil stiffness and damping. The estimation of stiffness and damping characteristics of cyclically loaded soil should be done at low (<0.001%) and high (>0.001%) shear strains. Some of the method to evaluate the dynamic soil parameters is given in Table 6.3. An *in situ* test of strain levels, which aims at making field measurements compatible with laboratory technique, is described in [5].

6.2.1.1 Group A

For low-level strain, where soil behavior is in the elastic range, the procedures to evaluate elastic stiffness are

1. Geophysical exploration from the ground surface: reflection and refraction method and surface wave technique
2. Geophysical borehole logging: PS-suspension, DH-downhole and CH-crosshole methods
3. Seismic cone penetration test: downhole and crosshole procedures

The value of shear modulus obtained through these tests provides benchmark stiffness or initial or maximum shear modulus (G_{max}).

6.2.1.2 Group B

For moderate strain and pre-failure behavior a pressure meter test and plate load test are applied.

Table 6.3 Field and laboratory test performed to predict dynamic soil property

Field test		Laboratory test	
Low strain test	High strain test	Low strain test	High strain test
Seismic reflection	Standard penetration test	Resonant column test	Cyclic triaxial test
Seismic refraction	Cone penetration test	Ultrasonic pulse test	Cyclic direct simple shear tests
Suspension logging	Dilatometer test	Piezoelectric bender element test	Cyclic torsional shear test
Spectral analysis of surface wave	Pressuremeter test		Shaking table test
Seismic cross hole test			Centrifuge test
Seismic cone test			

In this case, stiffness is associated with certain strain level. The shear modulus must be calculated based on plasticity theory.

6.2.1.3 Group C

For high-level strain, characterizing failure condition or even residual condition.

Pressure meter and plate load tests up to failure conditions and the Standard Penetration Test as well as the Cone Penetration Test can give this information. The stiffness value obtained through these tests is not related to the specific strain level. These test results should be empirically correlated with the reference stiffness given by Group A and Group B procedures (Figure 6.3).

6.2.1.4 *Following consideration should be made to determine* in situ *dynamic properties of soil*

- Representativeness of the sample of soil mass if it is stratified, nonhomogeneous
- Level and effect of soil disturbance while taking out the soil sample
- *In situ* condition and strain range of interest from the soil sample for particular problem

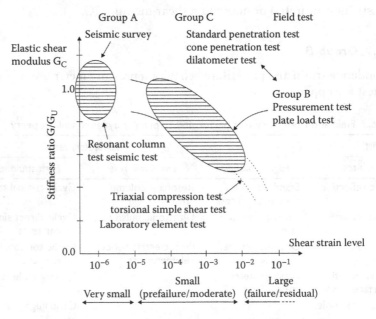

Figure 6.3 Comparative study of nonlinear shear deformation. (From Burland, J.B., *Proceedings of 1st IS-Hokkaido '94*, 2, 703–705, 1995.)

- Driving system compliance and possible bedding effect in laboratory equipment that can provide the true deformation characteristics of soil specimen

6.2.1.5 Comparison of laboratory and field test results

Laboratory and field test techniques are not ideal. Soil conditions can be better controlled in the laboratory, as field conditions do not permit adequate control of some of the factors influencing the stiffness characteristics of the soil. Correct interpretation and comparison of field test results also require laboratory tests so that the sensitivity of various parameters can be judged under repeatable conditions. It is known that laboratory test results are affected to some degree by the sampling, handling, and preparation methods used. In such cases, laboratory shear modulus to field modulus values can be less than or more than 1. It is also seen that medium-to-hard, over-consolidated soils are most sensitive to disturbance (Figure 6.4). Table 6.3 classifies different types of laboratory and field dynamic tests to quantify shear modulus.

6.2.1.6 Stress–strain behavior of cyclically loaded soil

The behavior of soil is quite complex, even in a static scenario, and so it becomes challenging when dynamic behavior is taken into consideration.

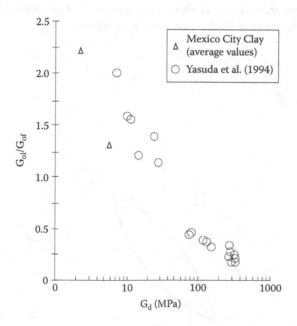

Figure 6.4 Soil stiffness influence on lab and field test result (G_{ol}/G_{of}). (From Yasuda et al., 1st IS-Hokkaido '94, 1, 197–202, 1994.)

The correct prediction of stress–strain behavior is very important for accurate modeling of soil. The dynamic property of soil depends on the level of strain induced; hence in the evaluation of shear modulus of soil, it is important to define the statin at which the modulus is obtained. The stress–strain behavior is broadly classified into equivalent linear model, cyclic nonlinear model, and advanced constitutive model. Out of these, the simplest one is the equivalent linear model, but it has limited ability to represent the actual behavior of soft and stiff clays that show nonlinear behavior. The behavior of soil is nonlinear, inelastic under cyclic loading conditions. The stiffness of soil is greatest and damping is least at low strain level. At high-strain level, the effect of soil nonlinearity increases, which produces greater damping and lower stiffness. Other advanced constitutive models can be used to simulate the nonlinear behavior of soil but there are many complexities.

6.2.1.3.1 Equivalent linear model

When cyclic loading is applied on to the soil, loading and unloading patterns can be described by a hysteresis loop. This model treats soil as a linear viscoelastic model. The shape and inclination of the loop are two important properties. Inclination depends upon the loading pattern, which is defined by a tangent modulus that varies throughout the cycle, but an average of G_{tan} is approximated by secant shear modulus, which describes the inclination of the hysteresis loop, as shown in Figure 6.5. The secant shear modulus (G_{sec}) is the ratio of shear stress to shear strain:

$$G_{sec} = \frac{\tau}{\gamma} \tag{6.1}$$

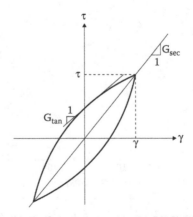

Figure 6.5 Hysteresis loop showing G_{sec} and G_{tan}.

where:

G_{sec} = secant shear modulus

τ = shear stress

γ = shear strain

The breadth of the loop is indicated by the area, which is the measure of energy dissipation while loading and reloading and is denoted by the damping ratio, shows how the shear modulus decreases at large strain (Figure 6.5). The damping ratio is given by

$$\xi = \frac{W_d}{4\pi W_S} = \frac{1}{2\pi}\left(\frac{A_{Loop}}{G_{sec}\gamma^2}\right) \tag{6.2}$$

where:

W_d = dissipated energy

W_S = maximum strain energy

A_{Loop} = area of hysteresis loop

G_{sec} = secant shear modulus

ξ = damping ratio

6.2.1.3.2 Cyclic nonlinear model

The accurate representation of the nonlinear model is carried out by cyclic nonlinearity that follows actual stress-strain paths during cyclic loading based on backbone curve, loading and unloading behavior, and stiffness degradation curve. In contrast to the equivalent linear model, the cyclic nonlinear model allows permanent strain to develop. To add complexity to the model, soil densification, irregular loading, and pore pressure generation should be added. The applicability of the model is dependent on the phenomenon of critical state soil mechanics, which point toward the stress path.

The shape of backbone curve can be described by hyperbola, dependent on initial stiffness (low strain) and shear strength of soil. The simplest hyperbolic equation to represent the variation of shear stress to shear strain is given by

$$\tau = \frac{G_{max}\gamma}{1+\left(\dfrac{G_{max}}{\tau_{max}}\right)\gamma} \tag{6.3}$$

6.2.1.3.3 Advanced constitutive model

The most accurate method to represent soil behavior is through the advanced constitutive model, which uses concepts of mechanics to explain soil behavior for initial stress condition, stress path, cyclic or monotonic loading, low- or high-strain rates, and drained and undrained conditions.

Soil models generally require the following parameters to describe accurate soil behavior:

1. Yield surface: to capture elastic behavior of soil.
2. Hardening law: explains the shape and size of yield surface as plastic deformation occurs.
3. Flow rule: describes increment of plastic strain with respect to increment to stress.

Some of the advanced constitutive soil models are cam clay [7], modified cam clay [8], and hyperbolic [9].

Advanced constitutive models allow flexibility in modeling to capture the response of soil to cyclic loading, but it requires more parameters than the equivalent linear or cyclic nonlinear model.

6.2.1.3.4 Selection of dynamic soil parameters

Shear wave velocity and shear modulus of soil at low strain are very important parameters in the analysis of earthquake problems. Also, shear modulus is the function of amplitude of shear strain under cyclic loading. Strain dependent modulus reduction and damping curves are generated from cyclic tri-axial tests.

6.2.1.3.4.1 SHEAR MODULUS

Soil stiffness is influenced by cyclic strain amplitude, void ratio, mean principal effective stress, plasticity characteristics, overconsolidation ratio, and number of loading cycles. The secant modulus of soil varies with cyclic shear strain amplitude. The secant shear modulus is high at low strain but decreases as strain amplitude increases. The locus of points corresponding to the tips of the hysteresis loops having various cyclic strain amplitudes are used to construct backbone curve, as shown in Figure 6.6.

6.2.1.3.4.2 MAXIMUM SHEAR MODULUS

Most geophysical tests are carried out at low shear strain (0.0003%), which measures the shear velocity of soil and can be used to compute the G_{max}:

$$G_{max} = \rho V_s^2 \tag{6.4}$$

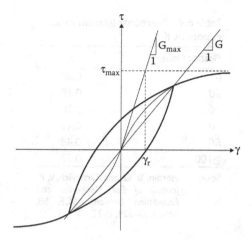

Figure 6.6 Backbone curve of shear stress vs. shear strain.

where:

ρ = density of soil

V_s = shear wave velocity of soil

G_{max} can also be evaluated from the laboratory test, which is given by

$$G_{max} = 625\, f(e)\left(OCR\right)^K P_a^{1-n}\left(\sigma'_m\right)^n \tag{6.5}$$

where:

$$f(e) = \frac{1}{0.3 + 0.7e^2}\,[11]$$

$$f(e) = \frac{1}{e^{1.3}}\,[12]$$

$$\sigma'_m = \frac{\sigma'_1 + \sigma'_2 + \sigma'_3}{3} = \text{principle effective stress}$$

$OCR = $ overconsolidation ratio

The overconsolidation ratio exponent (K) is shown in Table 6.4.

6.2.1.3.4.3 MODULUS REDUCTION CURVE (g/g$_{max}$) AND DAMPING RATIO (ξ)

The variation of modular ratio with shear strain is described graphically with the modulus reduction curve, as shown in Figure 6.7. Modulus reduction curves for coarse- and fine-grained soils are given by [13]. Refs. [14]

Table 6.4 Overconsolidation ratio exponent K

Plasticity index	K
0	0.00
20	0.18
40	0.30
60	0.41
80	0.48
≥100	0.50

Source: Hardin, B. O. and Drnevich, V. P., *Journal of Soil Mechanics and Foundation Division*, ASCE, 98, SM6, 603–624, 1972.

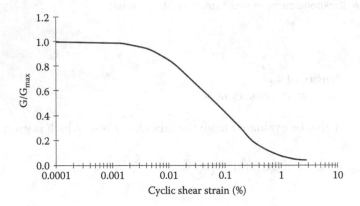

Figure 6.7 Shear modulus reduction curve with cyclic shear strain.

and [15] first noted the influence of soil plasticity on the shape of curve; shear modulus of high-plasticity soil was observed to degrade slowly with shear strain. The results of [16,17] show that the modulus reduction curve is influenced by the plasticity index and void ratio.

For many geotechnical investigations, dynamic *in situ* properties are not evaluated due to high cost; instead, the shear wave velocity of soil is evaluated based on well-established correlations with SPT N-value and soil indices, which are widely used (Table 6.5).

Damping behavior is influenced by plasticity characteristics of soil [15–17]. The damping ratio of high-plasticity soil is lower than that of low-plasticity soil at the same cyclic strain amplitude. Damping behavior is also influenced by effective confining pressure for the soil of low plasticity, and it increases with increasing strain amplitude.

Table 6.5 Empirical relationship between G_{max} and *in situ* test procedure

In situ test	*Relationship*	*Soil type*	*Comments*
SPT	$G_{max} = 20{,}000\{N_1\}_{60}^{0.333}(\sigma'_m)^{0.5}$	Sand	G_{max} and σ'_m) in lb/ff^2
	$G_{max} = 325\,N_{60}^{0.68}$	Sand	G_{max} in $kips/ft^2$
CPT	$G_{max} = 1634\,\{q_c\}^{0.250}(\sigma'_v)^{0.375}$	Quartz sand	G_{max}, q_c and σ'_m in kPa; based on field tests in Italy and on calibration tests
	$G_{max} = 406\,\{q_c\}^{0.695}e^{-1.1.30}$	Clay	G_{max}, q_c and σ'_m in kPa; based on field tests at worldwide sites
DMT	$G_{max}/E_d = 2.27 \pm 0.59$	Sand	Based on calibration chamber tests
	$G_{max}/E_d = 2.2 \pm 0.7$	Sand	Based on field tests
	$G_{max} = 530/[(\sigma'_v/p_a)^{0.25}] \cdot [Y_D/Y_w - I]/$ $[2.7 - Y_D/Y_w] \cdot K_0^{0.25}(p_a\sigma'_v)^{0.5}$	sand, silt, clay	G_{max}, p_a, σ'_m in same units; Y_D is dilatometer-based unit weight of soil; based on field tests
PMT	$3.6 \le G_{max}/G_{ur,c} \le 4.8$	Sand	$G_{ur,c}$ is corrected unloading-reloading modulus from cyclic PMT
	$G_{max} = 1.68/\alpha_p G_{ur}$	Sand	G_{ur} is secant modulus of unloading-reloading potion of PMG; α_p is factor that depends on unloading-reloading stress conditions; based on theory and field test data

6.3 FREE-FIELD GROUND RESPONSE ANALYSIS

The damage patterns of many recent earthquakes around the world, including the 1999 Chamoli and 2001 Bhuj earthquakes in India, have demonstrated that the effect of local soil condition on the level of ground shaking cannot be ruled out. For example, the epicenter of the 1985 Mexico City earthquake was located off the Pacific coast, but the damage occurred 360 km away. Similarly, the city of Delhi, which is 250 km away from the epicenter of the Chamoli earthquake, experienced moderate damage to structures overlying soft soils. The severe damage observed during the 2001 Bhuj earthquake also destroyed Ahmadabad city, which was located 250 km away from the epicenter, due to amplification of ground motion through soft alluvium. Figure 6.8 shows collapsed buildings in Ahmadabad after the earthquake.

Free-field ground motions are the motions that are not affected by the presence of structure. Analyses of these involve identification of potentially active sources in the region, estimation of seismicity associated with

Figure 6.8 View of collapsed buildings in Ahmadabad after the 2001 Bhuj earthquake in India.

individual sources, estimation of travel path influencing characteristics as they propagate from source to site, computation of dynamic response of soil deposit and soil–structure interaction, and assessment of stability under designed seismic excitation. They also involve modeling of fault rupture at the source of an earthquake, of seismic wave propagation through the earth to top of bedrock, and then of the passing of bedrock motion through the soil lying above it. The small travel distance of seismic waves from bedrock to topsoil has tremendous potential to influence the soil's characteristics and its response depend on the frequency content and duration of earthquake motion, geometry and material property of soil deposits lying above the bedrock. Prediction of the site-specific dynamic response of layered soil deposit in the region where earthquake hazard exists is a challenging task for geotechnical practitioners. Hence, in real practice, fault rupture mechanism is so complicated that the ground response analysis problem is reduced to determining the response of the soil deposit to the motion of the bedrock immediately beneath it. The response of level or gently sloping soil sites with horizontal layers and nearly vertically propagating shear waves can be obtained by one-dimensional ground response analysis and its results in terms of acceleration, displacement, stress–strain time histories, and design response spectra are widely used to evaluate liquefaction potential and to determine earthquake-induced forces that cause instability to structures. However, for sloping or irregular ground surfaces, the presence of heavy structures or stiff, embedded structures, or walls and tunnels requires two- or three-dimensional analyses [4].

The dynamic response of a structure depends on the type of supporting soil. During an earthquake event, deformations in the soil can be imposed onto the foundation. This further influences the ground motion of the superstructure, which makes the structural and ground deformations interdependent and can be referred to as soil–structure interaction. The supporting soil has potential to modify the response of the structure. The response depends on the thickness of soil above bedrock and characteristics of the ground motion passing through it. The fundamental period of soil deposit and the fundamental period of vibration of structure are of great significance, and their close match may create a condition of resonance. This must be taken into account when assessing the dynamic response of many structures and foundation systems.

6.3.1 Parameters influencing ground response analysis

Ground response analysis primarily depends on the following factors [4]:

- Mechanism of fault rupture at the source of earthquake.
- Stress wave propagation through crust to the top of bedrock.
- Local soil condition above bedrock.

Prediction of ground response is complicated by the uncertainties involved, such as the nature of energy transmission between source and site, accurate prediction of crustal velocities, and damping characteristics and mechanism of fault rupture. The solutions to the above problems may be overcome by performing a seismic hazard analysis (deterministic or probabilistic) for predicting the bedrock motion at the site location. These analyses are primarily dependent on attenuation relationship to predict the bedrock motion parameters. Ground response analysis may then be performed once the underlying bedrock motion is evaluated.

To evaluate the effect of local response, it is important to carry out geotechnical investigations and laboratory testing and to obtain a shear wave velocity profile along the soil depth. A modulus reduction curve and damping ratio curve that represent the variation of shear modulus (G/G_{max}) and damping ratio with strain may be obtained through a cyclic tri-axial test. Parameters necessary for ground response analysis are shown in Table 6.6.

6.3.1.1 Main factors that influence local site effect

- Seismological: intensity, frequency, content, and duration of bedrock motions
- Geological: local geologic structure, rock type, soil deposit thickness, stratigraphic characteristics, soil types
- Geotechnical: elastic vibration characteristics, impedance variation between bedrock and overlying soil, nonlinear behavior of soils, including fatigue-type effects of shaking duration

Table 6.6 Consideration of parameters for ground response analysis

Geotechnical characteristic of soil profile	Time-history of Input bedrock motion	Dynamic soil properties
• It involves number of soil layers and its thickness • Soil type • Initial damping ratio • Unit weight of soil • Shear modulus or shear wave velocity	• Time history of input motion can be taken from recorded accelerogram matching to regional seismo-tectonic characteristics • Synthetic and artificial bedrock motion obtained from scaling of recorded strong ground motion can be used	• Defined by the means of damping ratio and shear modulus reduction curves • Curves of variation of equivalent damping ratio and secant shear modulus (G_{sec}) with cyclic shear strain • Selected on the basis of type of soil, unit weight and plasticity characteristics of soil

• Geometrical: non-horizontal soil deposit, topography of bedrock underlying soil deposits, basin configuration

6.3.2 Wave propagation and site amplification

The ground motion such as the seismic waves propagates through overlying soil and reaches the ground surface. Local soil condition plays an important role in modifying the ground motion. This is known as *soil amplification*. The influence of ground deposit on bedrock movement depends on seismological impact, geological conditions, and the geotechnical characteristics of the site. The physical aspect of the problem is to predict the characteristics of the seismic motion along soil stratum. Seismic excitation at one location cannot be felt instantly at another location; it takes some time for the effects of excitation to be felt at different locations, and the effects are dependent upon the stiffness and attenuation characteristics of the medium. Generally, the geological materials are treated as continua and the dynamic response of these materials to dynamic or transient loading such as earthquakes, blasts, traffic-induced vibrations, and so on, is evaluated in considering one- or two-dimensional wave propagations mainly based on their geometrical and loading characteristics [27].

6.3.3 Assumptions of analysis

The techniques of ground response analysis are grouped as one-dimensional (like DEEPSOIL, SHAKE 2000, or EERA), two-dimensional (like VERSAT2D or FEM codes) and three-dimensional (like FLAC3D or

PLAXIS3D) according to the problem required to be solved. Simplified models based on one-dimensional wave propagation analysis are widely used for its simplicity and fair approximation of the complexities involved in the analysis. The following assumptions are considered in one-dimensional analysis.

1. The soil layer is considered as horizontal and semi-infinite.
2. The ground surface is level.
3. The incident earthquake waves are uniform, horizontally polarized shear waves and propagate in vertical directions.

6.3.4 Different approaches for free-field ground response analysis

6.3.4.1 Linear approach

The linear approach of ground response analysis involves derivation of closed form analytical solutions for idealized geometries and soil properties. It requires constant value of shear modulus (G) and damping ratio (ξ) for induced level of shear strain in each layer. However, soil behavior is inelastic, and its material properties change in space. Thus, a numerical technique is the appropriate option for ground response studies [28]. Nonlinear, hysteretic, stress-strain properties of soil are shown by using an equivalent linear method of analysis.

6.3.4.2 Equivalent linear approach

The equivalent-linear method of analysis is an iterative approach and is widely used to approximate nonlinear response of soil under seismic excitations through nonlinear hysteretic stress-strain behavior by using equivalent linear shear modulus G_{sec} and the equivalent linear damping ratio ξ. The variation of G and damping ratio "ζ" with cyclic shear strain is represented by modulus reduction curves and damping curves. The secant shear modulus G and damping ratio ξ produce the same energy loss as the actual hysteretic loop in a single cycle. Here, soil profile with bottom layer as half-space and extended to infinity in all directions, considers the response of soil deposits mainly due to vertical propagation of waves.

For the equivalent linear method, an iterative procedure is used to obtain the values of shear modulus and damping, compatible with the representative effective shear strain in each layer. The iterative procedure works as follows [4]:

1. Initially shear modulus $G^{(1)}$ and damping ratio $\xi^{(1)}$ are estimated for each layer corresponding to small strain level. These values are taken from the modulus ratio and damping curves.

2. The ground response, including the shear strain time history, is calculated by using $G^{(1)}$ and $\xi^{(1)}$ for each layer. The values of maximum shear strain $\gamma_{(1)max}$ are obtained from the shear strain time history for each layer.
3. The effective shear strain $\gamma^{(1)}_{eff}$ in each layer is estimated from maximum shear strain (taken approximately as 65% of the maximum strain). For layer "m," $\gamma^{(1)}_{eff\ (m)} = R_\gamma\ \gamma^{(1)}_{max(m)}$, where R_γ is the ratio of the effective shear strain to maximum shear strain and depends on the earthquake magnitude (Equation 6.6):

$$R_\gamma = \frac{(M-1)}{10} \tag{6.6}$$

4. The new equivalent linear values of the next iteration, $G^{(2)}$ and $\xi^{(2)}$ are selected corresponding to the calculated $\gamma^{(1)}_{eff}$.
5. Steps 2 to 4 are repeated until the computed values of G and ξ are obtained in two successive iterations that are nearly the same.

The equivalent linear method is thus an iterative method and approximates the nonlinear behavior of soil, by carrying out a piecewise linear analysis in each iteration.

6.3.4.3 Nonlinear approach

Although the simplicity of the equivalent linear approach makes it computationally convenient and provides reasonable answers to practical problems, it remains an approximation to the actual nonlinear response of soil. The tangent shear modulus at each strain level is considered in the nonlinear analysis, and constitutive models, which are usually cyclic stress-strain models, are used to represent the stress-strain behavior of the soil. The analysis is carried out by using direct numerical integration at small intervals in the time domain. By integrating the equation of motion in small time step, any nonlinear model or advanced constitutive model may be used. The available nonlinear computer program characterizes the stress-strain behavior of soil by cyclic stress-strain models such as hyperbolic models, modified hyperbolic models, the Hardin-Drnevich-Cundall-Pyke (HDCP) model, and so on.

6.3.5 Steps to be followed for the free-field analysis

The ground response should be carried out in the following steps:

- Selection of dynamic soil properties based on *in situ* laboratory test results and well-established empirical correlation. This includes the selection of a modulus reduction curve and a damping ratio curve and other parameters. More elaborate dynamic testing procedures are required for special problems.

- Determining the characteristics of the motion likely to develop in the rock formation underlying the site. The peak ground acceleration, predominant period, and effective duration are key parameters of any seismic motion and the empirical relationship between these three parameters and the causative fault to site distance should be established for magnitudes of earthquakes expected at the site [29,30,28,32,34].
- A design motion with the desired characteristics can be selected from strong motion accelerograms that have been recorded during previous earthquakes or from input earthquake motion through the Uniform Hazard Response Spectra (UHRS) selected from the provided Probabilistic Seismic Hazard Analysis (PSHA) for the desired return periods or GRA scaling and guidelines provided by the NEHRP, NBC, or ASCE.
- Synthetic earthquake time histories for DBE and MCE are generated using available computer programs based on selected seed accelerograms and target response spectra at the bedrock level. Other responses of GRA, like acceleration response, soil amplification response, variation of PGA along soil depth, and so on, can be obtained for various damping values and can be used for design.

Many researchers have adopted this procedure to obtain the response of subsoil during different earthquakes. An equivalent and nonlinear approach to obtain seismic ground response was used in [31] for Goa, India.

6.4 LIQUEFACTION OF SOIL

6.4.1 Introduction

Liquefaction is a phenomenon in which loose, non-cohesive sand and silt below water table level develop high pore water pressure, which reduces the shear strength of the soil mass causing the soil to behave as fluid during a strong earthquake. Its devastating effects were observed in the 1964 Good Friday earthquake in Alaska and in the 1964 Niigata earthquake in Japan. Extensive damage occurred to multistoreyed buildings, bridge foundations, and buried structures. Figure 6.9 shows liquefaction failures that occurred to buildings in the Niigata earthquake. Figure 6.10 shows liquefaction failures during the 1989 Loma Prieta earthquake in the USA. Recently reclaimed lands along the coastline are highly susceptible to liquefaction, and several port facilities were significantly damaged by liquefaction-induced ground displacements during 1995 Kobe earthquake in Japan. In the 2001 Bhuj earthquake in India, liquefaction failures were characterized by widespread sand boils, craters and lateral spreading, extensional cracks, monoclinal folds, and tear faults [33].

Figure 6.9 Tilting of apartment buildings at Kawagishi-Cho, Niigata during the 1964 Niigata Earthquake due to soil liquefaction. (From Kramer, S.L., *Geotechnical Earthquake Engineering.* Prentice-Hall, Upper Saddle River, NJ, 1996.)

Figure 6.10 Liquefaction in recent deposits of the Parajo River during 1989 Loma Prieta earthquake in California.

Increased pore water pressure reduces the shear strength and stiffness of the soil deposit. The increase in pore water pressure causes an upward flow of water to the ground surface, where it emerges in the form of mud spouts or sand boils [22]. The dissipation of excess pore water pressure starts after the liquefaction. The time duration up to which soil will remain in the state of liquefaction depends on two major factors: (1) the duration of the seismic shaking and (2) the drainage pattern in the liquefied soil. It can be stated that the longer and the stronger the cyclic shear stress application exists, the longer the state of liquefaction persists. Similarly, if the confinement of a liquefied layer is by clay from both the upper and the lower side, then time of dissipation of excess pore water pressure generated will be longer. The liquefaction phenomenon can alter the natural state of soil, which may change from loose to dense.

Soil liquefaction depends on the magnitude of earthquake, intensity and duration of ground motion, the distance from the source of the earthquake, site-specific conditions, ground acceleration, type and thickness of the soil deposit, relative density, grain size distribution, fines content, plasticity of fines, degree of saturation, confining pressure, permeability characteristics of soil layer, position and fluctuations of the groundwater table, reduction of effective stress, and shear modulus degradation [4,35–37].

6.4.2 Evaluation of liquefaction potential of soil

Liquefaction potential evaluation has been attempted by a host of methods/techniques by various researchers. These include use of "simplified procedures" based on standard penetration tests (SPT), cone penetration tests (CPT) and shear wave velocity (Vs) criteria developed from field liquefaction performance cases histories. These methods are also known as deterministic methods, in which liquefaction of a soil is predicted to occur if the factor of safety (FS) [defined as the ratio of cyclic resistance ratio (CRR) over cyclic stress ratio (CSR)] is less than or equal to 1. No soil liquefaction is predicted if the FS is more than 1. However, in reality, it is well understood that this boundary is not distinct, and hence engineering judgment needs to be applied to evaluate the liquefaction susceptibility of soils with an FS near 1. Furthermore, there are several possible procedures that may be employed for each *in situ* test method (i.e., SPT, CPT, Vs) and the FS obtained from each of the procedures are not equivalent for the same input data. Therefore, caution must be exercised when establishing an FS, and it is advisable to check the results from more than one procedure.

The majority of liquefaction-related studies concentrate on relatively clean sands. It was previously believed that only "clean sandy soils" with few fines would liquefy, and cohesive soils were considered to be resistant to cyclic loading. However, the 1975 Haicheng and 1976 Tangshan earthquakes showed that even cohesive soils could liquefy. Also, recent ground failure case histories after the 1994 Northridge, 1999 Kocaeli, and 1999 Chi-Chi earthquakes further illustrated that silty and clayey soils may be susceptible to soil liquefaction. This has accelerated research studies on the liquefaction susceptibility of fine-grained soils.

The first notable effort to identify potentially liquefiable fine-grained soils was the Chinese Criteria [38]. This criterion was used with some modifications until the case histories compiled after some recent earthquakes

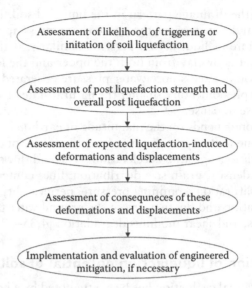

Figure 6.11 Key Elements of Soil Liquefaction Engineering. (From Schnabel et al., SHAKE: A computer program for earthquake response analysis of horizontally layered sites. EERC Report 72-12. Earthquake Engineering Research Center, Berkeley, California, 1972.)

like the 1994 Northridge, 1999 Kocaeli, and 1999 Chi-Chi earthquakes verified that neither the original form, nor the modifications can successfully identify soils liquefied during and after these earthquakes. This fact has increased research focusing on the development of new criteria based on field observations and the results of laboratory cyclic tests on "undisturbed" samples from liquefied sites. New criteria were proposed [39–42,56] for the assessment of the liquefaction potential of fine-grained soils recently based on the Plasticity Index (PI). PI can confidently distinguish between fine-grained soil behavior that is "clay-like", for which cyclic softening is expected, or "sand-like", which is liquefiable.

Figure 6.11 summarizes major components of soil liquefaction engineering [40]. As illustrated, the starting point of liquefaction engineering is to decide whether soil is susceptible to liquefaction or not, since assessment of the likelihood of liquefaction would be meaningless for non-liquefiable soils.

6.4.3 Liquefaction susceptibility criteria

Liquefaction susceptibility criteria decide whether the *in situ* soil is susceptible to liquefaction or not. The prediction of which is based on *in situ* tests, laboratory results, and empirical correlations.

6.4.4 Simplified approaches for estimating liquefaction potential of cohesionless soils based on standard penetration test (SPT)

Different methods for assessments of liquefaction potential were developed by various researchers [43–46]. In the recent past, simplified procedures for the evaluation of liquefaction potential were proposed by [47,37,48–51,53]. These methods, also known as deterministic methods, were based on standard penetration tests (SPT), cone penetration tests (CPT) and shear wave velocity (Vs) criteria and were developed from field liquefaction performance cases at sites that had been characterized with the corresponding *in situ* tests [43].

The most commonly used technique is SPT. The results, which are commonly referred to in terms of N-value, follow certain protocols:

1. Estimation of the cyclic stress ratio (CSR) generated along the soil depth by seismic shaking.
2. Estimation of the cyclic resistance ratio (CRR) of the soil that is required to cause initial liquefaction in the soil.
3. Evaluation of factor of safety (FS) against liquefaction potential of *in situ* soils by dividing the cyclic stress that *in situ* soil can withstand without liquefying (CRR) by the stress induced by the seismic shaking (CSR), according to Equation 6.7:

$$FS = \frac{CRR}{CSR} \tag{6.7}$$

6.4.4.1 Evaluation of cyclic stress ratio (CSR)

The simplified procedure was developed by [43] as an easy approach to estimate earthquake-induced stresses without the need for a site response analysis and has been the primary method to determine the CSR since early 1970s. The simplified CSR is calculated according to

$$CSR = \frac{\tau_{av}}{\sigma'_v} = 0.65 \cdot \left(\frac{a_{max}}{g}\right) \cdot \left(\frac{\sigma_v}{\sigma'_v}\right) \cdot r_d \tag{6.8}$$

where:

a_{max} is the peak horizontal ground surface acceleration modified for site-specific soil conditions

g is the acceleration due to gravity

σ_v is the total overburden stress

σ'_v is the initial effective overburden stress

r_d is a stress reduction coefficient that takes the flexibility of the soil column into account.

All simplified liquefaction evaluation procedures use this common equation to determine the CSR. However, the various procedures often use different relationships to determine the coefficient r_d. As such, differences in the CSR calculations between various simplified methods are primarily a direct result of the uncertainty in determining r_d.

6.4.4.2 Evaluation of cyclic resistance ratio (CRR)

The CRR is the dynamic stress the soil can withstand before liquefying. SPT-based liquefaction triggering procedures use blow count (N) as the basis of computing CRR. The SPT-based CRR relationships are well-established procedures for evaluating CRR. CRR equations for each procedure are presented as $CRR_{7.5,1\,atm}$ (the cyclic resistance ratio of the soil adjusted to 1 atmosphere of effective overburden pressure for a $M_w = 7.5$ earthquake). This is necessary because the various relationships use different factors to account for earthquake magnitude and overburden stresses. It is common for CRR relationships to be normalized for this base case and then site-specific adjustments are made via magnitude scaling factors (MSF) and overburden correction factors (K_σ) to account for the earthquake magnitude under consideration and the overburden stresses at the depth of interest. The site-adjusted CRR is obtained from $CRR_{7.5,1\,atm}$ according to

$$CRR_{M,K\sigma} = CRR_{M=7.5,1\,atm} \cdot MSF \cdot K\sigma \qquad (6.9)$$

where all variables are defined as above.

Though all the simplified procedures use a common equation to determine the CSR based on the seismic (ground motion) parameters derived from the ground response analysis for specific site, CRR varies significantly between these procedures because each procedure use a different equation for determination of CRR for the base case and different factors to account for earthquake magnitude and overburden stresses.

6.4.4.3 Evaluation of liquefaction potential or cyclic failure of silts and clays

Clay, on the other hand, behaves quite differently. Clays are compressible where sands generally have a small compressibility. The void ratio or density of clay is dependent upon effective consolidation stress and stress history rather than depositional environment. Using high-quality field sampling, laboratory testing of clays can provide reliable test results to evaluate cyclic loading behavior. *In situ* tests can also provide qualitative information on the undrained shear strength of clay.

The first method to evaluate cyclic failure potential of soils that behave like clays was provided by [57]. New criteria was presented by the authors for distinguishing between fine-grained soils that will exhibit sand-like versus clay-like behavior during the undrained cyclic loading imposed by the earthquakes. A PI > 7 for clay-like behavior for fine-grained soils with a slight lower transition point for CL-ML soils (PI ≥ 5 or 6) is proposed in [44], as shown in Figure 6.12.

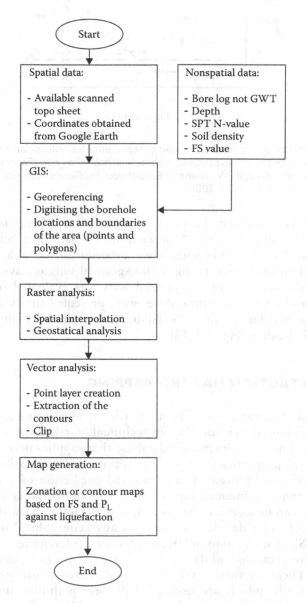

Figure 6.12 Flow chart for generation of GIS map for liquefaction hazard mapping.

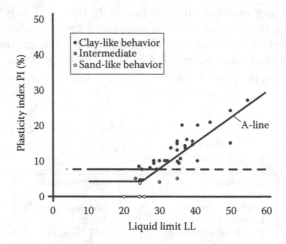

Figure 6.13 Atterbergs limits chart showing representative values for each soil that exhibited clay-like, sand-like an intermediate behaviour. (From Boulanger, R. W., and Idriss, R. W., *Journal of Geotechnical and Geoenvironmental Engineering.*, 132(11), 1413–1424, 2006.)

Failure of sand and clay during cyclic loading was differentiated in this paper [44] by using the term "liquefaction" for sand-like behavior and "cyclic failure" for Clay-like soils. These procedures were able to account for observed failure by evaluating the fine-grained soils as clays.

The seismic behavior of fine-grained soils was evaluated by [42] and guidelines and engineering procedures were presented for evaluating the potential for liquefaction or cyclic failure of low-plasticity silts and clays during cyclic loading (Figure 6.13).

6.5 LIQUEFACTION HAZARD MAPPING

Mapping has become a standard approach for identifying zones of liquefaction potential within the geotechnical earthquake engineering community. The geo-information database that supplies necessary information of soil properties for liquefaction potential calculation and hazard mapping can be used for analysis and prediction of geo-disasters such as earthquake-induced liquefaction. The vulnerability to liquefaction hazard can be assessed through these liquefaction hazard maps and the risk can be calculated. The ability of any geographical information system (GIS) to store, manipulate, display, and interchange or integrate the extensive geotechnical data has made its use not just convenient but necessary. There are various GIS-based applications to analyze and map seismic hazards, which are useful for disaster planning and management, and vulnerability and risk assessment. Additionally, the inclusion of two-dimensional and three-dimensional GIS is changing the way the seismic hazards are analyzed.

6.5.1 Recent advances in liquefaction hazard mapping

A GIS-based liquefaction susceptibility map for Mumbai, India was developed in [54]. A seismic liquefaction hazard analysis was performed in [55], which also suggested the design guideline for the design of piles in liquefiable soil deposits. Ground damage hazard assessment such as liquefaction, landslide disaster, and mapping with the aid of geographical information systems (GIS) through the research carried out in the Wellington Region of New Zealand is illustrated in [57]. A borehole-based geological database for the assessment of liquefaction hazards in the Kathmandu Valley is developed in [58], and [59] produced liquefaction hazard maps based on both regional and local data locates zones of liquefiable materials within a geologic deposit. The relative susceptibility of Quaternary geologic deposits to earthquake-induced liquefaction near St. Louis, Missouri, are assessed in [60] and the result of the geologic and geotechnical characterization and quantitative analysis were presented on the GIS-based, 1:24000-scale, liquefaction susceptibility maps. The key geotechnical and geological features were identified by the authors, which indicated the locations most susceptible to lateral flow due to liquefaction. A database for the city of Adapazarı based on surface observations as well as field and laboratory test results using GIS methods was established, and based on this, it is now possible to query the groundwater level, soil type, organic content, SPN, undrained shear strength, allowable bearing capacity and liquefaction potential for the top 15 m depth [61]. A liquefaction hazard-screening tool for the California Department of Transportation (Caltrans) that is being applied in evaluating liquefaction hazard to approximately 13000 bridge sites in California is developed in [62]. A GIS-based liquefaction prediction system was developed by [63] to determine the ground liquefaction risk of the Kunming Basin, Yunnan, China, due to a rare earthquake. A methodology is presented in [64] that helps the decision-making process and improves the ability of mapping liquefaction prone areas, by using the ArcGIS's Geostatistical Analyst extension. In [65], hazard maps of the liquefaction susceptibility of areas in the city of Damietta were developed, and displayed using the GIS to show a spatial variability observed in LPI values and the presence of a thin layer of liquefiable soil even in profiles of low LPI value. Imported into [66] are the outputs from Artificial Neural Network (ANN)– and SPT–based liquefaction methods, used for soil liquefaction assessment, into ArcGIS to map the liquefaction resistance, seismic settlements, liquefaction potential, and liquefaction prediction for the Western Izmit Basin, Turkey. Screening-level liquefaction hazard maps for the Australia region were created in [67], corresponding to design level earthquake having annual probability of exceedance equal to 10%, 5%, and 2% in 50 years (equivalent to return periods of 500, 1000 and 2500 years, respectively) based on AS1170.4. Liquefaction potential maps using GIS software for the Eskisehir urban area, situated within the second degree earthquake region on the seismic hazard zonation map of Turkey were produced in [69], considering factors of safety calculated at different depth intervals, geological setting and ground water level for different acceleration levels. Also, [69] used GIS to obtain soil liquefaction hazard map of Satte city as a result of an earthquake depending on the ground conditions and soil behavior.

6.5.2 Generalized procedure for liquefaction hazard mapping

These steps should be followed to develop liquefaction hazards maps using GIS:

1. Input geological data, soil data, map data, ground conditions
2. Evaluate factors of safety and liquefaction potential at field test locations
3. Manipulate, analyze, interpret, and interpolate the data in GIS
4. Contour the data to produce a spatial representation of liquefaction

These steps are a general description of the approach to produce a liquefaction hazard map. GIS can be adapted to any size operation, and data can be incorporated at any scale from a single field. The efficiency of GIS in data-processing distinguishes it from other methods and broadens the scope of analysis but at the same time increases isolation with its increased utility.

The flow chart of the step-by-step procedure for generation of GIS maps is shown in Figure 6.12.

6.6 SEISMIC ANALYSIS OF SINGLE PILE

Pile foundations have wide applications in civil engineering and are most popular form of deep foundations used for both offshore and onshore structures for transferring vertical as well as lateral loads from the superstructure to the deeper strata, when the topsoil is either loose or soft or of a swelling type. Piles are long, slender columns that are either driven or bored (cast *in situ*). Driven piles are made up of variety of materials such as concrete, steel, timber, and so on, whereas bored piles are made up of concrete only. Local soil conditions and topography play significant roles in the amplification of ground motion and influence the choice of foundation system for any superstructure. Liquefaction of soil subjected to seismic loading is a governing factor affecting the performance of pile foundation in seismically vulnerable areas. Ground motions, free-field site response, and soil–pile interaction have a significant impact on the behavior of piles in areas prone to liquefaction.

The behavior of pile foundations under the impact of seismic forces can be characterized as complex soil–structure interaction, which has a significant impact on the behavior of piles that pass through soil that can experience liquefaction or cyclic mobility. More attention needs to be given to pile foundations, as floating piles passing through liquefiable soil layers undergo significant loss of shaft resistance. In case of end-bearing piles, excessive loads may be transferred to the end-bearing strata due to loss of shaft resistance, and thus the pile is subjected to higher bearing pressures. Also loss of shear strength in the liquefied zone will increase the effective length of the pile and thus the pile may fail by buckling if axial loads are predominant.

6.6.1 Types of pile foundation

Piles can be classified according to the mode of transfer of load and its use.

6.6.1.1 Classification based on the mode of transfer of load

1. End-bearing piles transfer loads through their bottom tip. Such piles act as columns and transmit the load through a weak material to a firm stratum below. If bedrock is located within a reasonable depth, piles can be extended to the rock. The ultimate capacity of the pile depends upon the bearing capacity of the rock.
2. Friction piles transfer the load through skin friction between the embedded surface of the pile and the surrounding soil. Friction piles are used when a hard stratum does not exist at a reasonable depth.
3. Combined end-bearing and friction piles are transfer loads by a combination of end bearing at the bottom of the piles and friction along the surface of the pile shafts. The total load carried by the pile is equal to the sum of the load carried by pile tip and the load carried by the skin friction.

6.6.1.2 Classification based on type of piles

1. Load-bearing piles are used to transfer the load of the structure to a suitable stratum by end bearing, by friction, or by both.
2. Compaction piles are driven in to the loose granular soils to increase the relative density. The bearing capacity of the soil is increased due to densification caused by vibrations.
3. Tension piles are in tension. These piles are used to anchor down structures subjected to hydrostatic uplift forces, or overturning forces.

4. Sheet piles form a continuous wall or bulkhead, which is used for retaining earth or water.
5. Fender piles are sheet piles that are used to protect waterfront structures from impact by shipping vessels.
6. Anchor piles are used to provide anchorage for anchored sheet piles. These piles provide resistance against horizontal pull for a sheet pile wall.

6.6.2 Failure mechanism of single pile

Failures of piled foundations have been observed in the majority of recent strong earthquakes. The failures of end-bearing piles in liquefiable areas during earthquakes are attributed to the effects of liquefaction-induced lateral spreading in most of the reported case histories by [70–73].

Three possible mechanisms of failure of single piles can be assumed under liquefiable soil conditions.

- Single piles carrying large axial loads and passing through loose, cohesionless soil and resting on hard rock
  - The seismic loading may generate cyclic shear stress, which increases the excess pore water pressure in a sandy layer. This leads to loss in the strength of soil mass, and it may be subjected to huge force, which may lead to bulking instability in the pile. The pile may then fail by forming a plastic hinge, as shown in Figure 6.14. The location of a plastic hinge will be the boundary between sand and rock because that region point will be subjected to maximum moment under buckling. This case is similar to the case of loading on the cantilever.

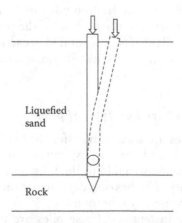

Figure 6.14 Mode of collapse of single pile subjected to buckling instability.

- Single pile carries large axial load and passes through a liquefiable loose, cohesionless layer resting on a non-liquefiable, dense sand layer
  - The loose, sandy layer will see the rise in pore water pressure under earthquake loading, which leads to degradation in the stiffness of soil mass and finally loss in the shear strength of soil mass. Excess pore pressure will be generated in the loose sand layer. Dense sand layers close to the interface between the loose and dense sand layer have reduced capacity, and excessive settlement may be noticed in such cases, as shown in Figure 6.15.
- Pile in sloping ground and passes through non-liquefiable sand layer and liquefiable sand layer both and finally resting on rock
  - A pile is passing through inclined and liquefiable sandy layer and overlain by non-liquefiable sandy layer, as shown in Figure 6.16. The induced liquefaction due to an earthquake event may lead to lateral spreading of the soil mass. The non-liquefiable layer will

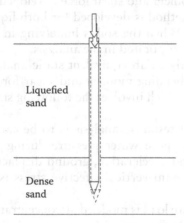

Figure 6.15 Mode of collapse of single pile subjected to bearing failure.

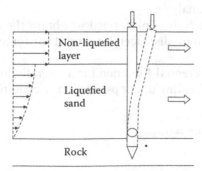

Figure 6.16 Failure of piles under combined lateral and axial load on laterally spreading soil with non-liquefiable layer.

act as a solid floating on fluid and impart very high lateral load on the pile, which may lead to buckling instability in the pile. The plastic hinge will form at the boundary of the rock socketing and the liquefiable sand layer.

6.6.3 Pseudo-static analysis of pile

Pile foundations need to be designed for carrying lateral loads in addition to vertical loads in foundations for high-rise buildings subjected to wind and seismic loads, quay and harbor structures subjected to lateral forces due to impact of ships during berthing and wave action, and offshore structures subjected to wave action. In the design of pile foundations, the ultimate and serviceability limit state need to be evaluated properly.

A simplified procedure to consider these effects is the pseudo-static approach in which an equivalent static analysis is carried out to obtain maximum bending moment and shear force developed in pile due to earthquake loading. This method is developed for both liquefying and non-liquefying soils [74–77]. When the soil is liquefying in nature, the stiffness degradation is to be incorporated in the analysis.

In pseudo-static analysis, an equivalent static analysis is carried out to obtain the maximum bending moment and shear force developed in piles due to lateral loading, which involves the following stages:

- Free-field ground response analysis is to be carried out considering the generation of pore water pressure during earthquake loading. Maximum ground acceleration, ground displacement along the soil depth, and minimum vertical effective stress is obtained from this analysis.
- The superstructure load is modeled as concentrated mass acting at the pile top to simplify the analysis.
- The horizontal force is applied at the pile head, which is equal to the product of pile cap mass and maximum ground acceleration obtained from free-field analysis.
- A nonlinear analysis is performed to obtain the maximum bending moment, maximum displacement, and shear force along pile length.

The governing differential Equation for a laterally loaded pile subjected to axial compressive loading under pseudo-static conditions is given by [78]

$$EI\frac{d^4y}{dz^4} + P\frac{d^2y}{dz^2} + ky = 0 \qquad (6.10)$$

where:
  y = lateral deflection of pile
  z = depth along the pile from top

EI = flexural rigidity of pile
P = pseudo-static load
k = ηz
η = modulus of subgrade reaction [kN/m³]

The general solution of above governing differential Equation is given by [78]

$$y = \left(C_1 e^{\beta z} + C_2 e^{\beta z}\right)\cos(\alpha z) + \left(C_2 e^{\beta z} + C_4 e^{\beta z}\right)\sin(\alpha z) \tag{6.11}$$

where:

$$\alpha = \sqrt{\sqrt{\frac{k}{4EI}} + \frac{P}{4EI}}$$

$$\beta = \sqrt{\sqrt{\frac{k}{4EI}} - \frac{P}{4EI}}$$

The above coefficients are valid for $P < 2\sqrt{kEI}$.

The integration constants C_1, C_2, C_3 and C_4 can be obtained by applying suitable boundary conditions at the pile head and pile tip, respectively. Finally, shear force, bending moment, and deflection along pile length can be obtained by using an iterative approach in a standard computer program in MATLAB, FORTRAN, and so on.

A pseudo-static analysis of free-headed single piles with floating tips using an analytical approach based on finite element procedure was carried out in [79]. The entire pile was discretized into various elements, each having height "h." So, the Equation 6.12 modifies to:

$$y = \left(C_1 e^{\beta \frac{z}{h}} + C_2 e^{\beta \frac{z}{h}}\right)\cos\alpha\frac{z}{h} + \left(C_3 e^{\beta \frac{z}{h}} + C_4 e^{\beta \frac{z}{h}}\right)\sin\alpha\frac{z}{h} \tag{6.12}$$

where:

$$\alpha = h\sqrt{\sqrt{\frac{k}{4EI}} + \frac{P}{4EI}}$$

$$\beta = h\sqrt{\sqrt{\frac{k}{4EI}} - \frac{P}{4EI}}$$

In the above equation, c_1, c_2, c_3 and c_4 are the unknown integrating constants, the values of which are determined after defining the appropriate boundary conditions at the various nodal points. The allowable vertical load acting at the pile top was computed after calculating the ultimate capacity of the pile in different types of soil and applying a factor of safety of 2.5 [80]. Five different ground motions were considered in the analysis: the 1989 Loma Gilroy, 1994 Northridge, 1995 Kobe, 2001 Bhuj, and 2011 Sikkim earthquake motions. Seismic equivalent linear ground response analysis was conducted using the SHAKE2000 computer program, and the maximum horizontal accelerations at various depths were obtained. The net lateral loads acting at various depths along the pile were computed after multiplying the allowable pile capacity with the maximum horizontal acceleration, and a "lateral load coefficient (a)," which is defined as the fraction of vertical load assumed to act at different depths along the pile length, was determined. The lateral load coefficient was considered to remain constant along the pile depth for a particular analysis, and parametric variation was conducted by varying it between 10% and 30% [80]. Nonlinear, pile–soil interaction analysis was conducted using MATLAB, and the influence of seismic motions, presence of vertical load at the pile top, lateral load coefficient, and variation in pile properties were studied.

It was also observed from ground response analysis that for the 1995 Kobe motion, while it traveled through the soil layers, the amplification of acceleration at the ground surface was comparatively less when compared to the 2001 Bhuj motion, although the former had higher acceleration at bedrock level. This may be attributed to the effect of soil nonlinearity and plasticity, which prevents a major increase in peak acceleration of an earthquake having higher acceleration at bedrock level. Moreover, the 2001 Bhuj motion had a higher frequency content and bracketed duration. This clearly indicates that the amplification of ground motions is significantly influenced by the soil type, duration, and frequency content of the input seismic motions, with the maximum horizontal acceleration having a lesser effect.

Figure 6.17 shows the variation of normalized bending moment with depth coefficient for a 20 m pile having a diameter of 500 mm and l/d ratio 40, embedded in dry, dense sand with $a = 0.3$ and $Q_v = Q_h$ and $Q_v = 0.0Q_h$, respectively. As observed from Figure 6.17, when the input motion changes from the 2001 Bhuj to 1995 Kobe earthquakes, the normalized moment rises by 40% while for the 2011 Sikkim, 1989 Loma Gilroy, and 1994 Northridge motions, the percentage increase is 2%, 15% and 33.3%, respectively, for $Q_v = Q_h$. However, for $Q_v = 0.0Q_h$, the percentage increase in normalized bending moment, is comparatively less, namely 1.6%, 12.1%, 21.4% and 28.6%, respectively for the 2011 Sikkim, 1989 Loma Gilroy, 1994 Northridge, and 1995 Kobe earthquake motions with respect to the 2001 Bhuj motion. It shows the influence of earthquake motions on pile behavior.

In Figure 6.18, the variation of normalized displacement with depth coefficient for the same pile embedded in dry, loose sand is illustrated. It is observed that when the input motion changes from 2001 Bhuj to 2011

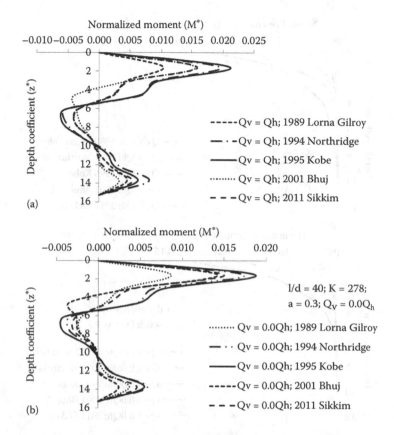

Figure 6.17 Variation of normalized moment (M*) with depth coefficient (z*) for a pile with l/d ratio 40 embedded in dry dense sand and (a) $Q_v = Q_h$ and (b) $Q_v = .0Q_h$. (From Chatterjee et al., *Computers and Geotechnics*, 67, 172–186, 2015.)

Sikkim, 1989 Loma Gilroy, 1994 Northridge, and 1995 Kobe, the normalized pile head displacement increases from 0.010 to 0.0154, 0.03, 0.08 and 0.126, respectively, for $Q_v = Q_h$ and a = 0.3. However, for $Q_v = 0.0Q_h$, the increase in normalized displacement is comparatively less. This shows the significance of considering the vertical seismic acceleration on pile behavior.

6.6.4 Dynamic forces on pile foundation

The predominant load on the pile foundation is vertical compressive load. Dynamic forces on piles may be due to liquefaction of layers of soil through which the pile foundation passes and may be due to dynamic loading onto the top of the pile foundation, due to railway loading (which may be due to high-speed trains), or to blast loading or strong seismic motions.

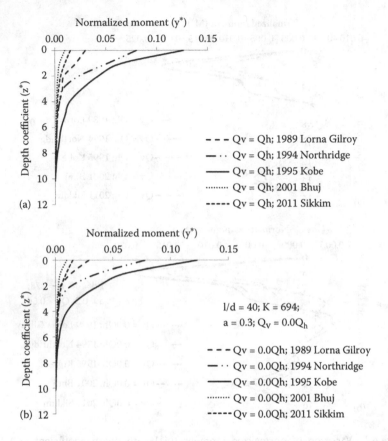

Figure 6.18 Variation of normalized displacement (y*) with depth coefficient (z*) for a pile with l/d ratio 40 embedded in dry loose sand and (a) Q$_v$=Q$_h$ and (b) Q$_v$=0.0Q$_h$.(From Chatterjee et al., *Computers and Geotechnics*, 67, 172–186, 2015.)

6.6.4.1 Liquefaction-induced forces on pile foundation

During seismic loading, liquefied soil tends to act as dense fluid and to move laterally. This is called lateral spreading, and it leads to a decrease in the strength of soil around the pile. This phenomenon leads to loss in shaft resistance in floating piles. Hence, the pile becomes unstable under axial load. Seismic loads of pile foundations are induced in the following circumstances.

- Inertial loading, which acts at the pile head and gets transferred from the superstructure to the pile. This force is usually cyclic in nature and is largely dependent on the frequency of input motion and superstructure. It varies as the earthquake persists and is normally at maximum value in the initial part of the shaking. This is before liquefaction, but once the soil liquefies, the force at the pile head reduces because of stiffness degradation of the soil.

- Kinematic loading acts along the length of the pile embedment to the depth of soil liquefaction and it is mainly due to the movement of the liquefied soil. The ground motion generates kinematic interaction between the piles. If the pile is embedded in a soft soil layer, then under the earthquake excitation, the soft layer will experience free-field displacement based on the natural frequency of the layer. This free-field displacement will be imposed on the pile foundation. If piles are relatively flexible, they will experience lateral deformation, but in the case of stiff piles, deformation will be much less, and hence the pile will be subjected to significant lateral load due to the mobilization of passive pressure in the opposite direction. In cases of sloping ground and where any non-liquefied crust exists above the liquefied soil layer, the extra force applied to the pile due to this crust will be considerably higher, which may be due to the down-slope movement under earthquake loading. The magnitude of the force depends upon the slope of the ground and the type of crust. Experimental investigations and a few case histories provide more insight into the concept behind the application of kinematic load during and after shaking, and sometimes as a combination of both. This force will be monotonic in nature if only kinematic force is acting (that is. the shaking has stopped), but it will normally be cyclic in nature if both kinematic and inertia forces both act together.

- Under cyclic loading, pore water pressure gets generated in the soil mass and its stiffness degrades, which in turn further degrades the shaft resistance of piles. The resistance to the pile for a particular amount of deformation is expressed as p–y soil springs. The stiffness of these springs degrades as soil liquefies. The results of a series of centrifuge modeling were applied in [81] to derive a correlation between stiffness degradation coefficient and the pore pressure ratio, r_u. In the case of full liquefaction, that is, 100% pore pressure, r_u reaches 1.0. These coefficients can be multiplied to the static p–y relations to obtain the dynamic p–u response.

- Combined effect of inertial and kinematic loading: [82] performed large-scale shaking table tests on pile groups in dry and saturated sands and obtained the phase dependency relation between the natural period of ground (T_g) and the structure period (T_b): (a) For $T_b > T_g$— Kinematic forces tend to be out of phase with inertial loads and peak pile stresses occur at an interim point (b) For $T_b < T_g$—Kinematic and inertial forces tend to be in phase and peak pile stresses occur when both effects are maximum.

In order to estimate the peak bending moment due to combined inertial and kinematic loads, [83,84] and [85] suggested the following method:

a. For $T_b > T_g$: The peak pile bending moment in the pile can be estimated by the SRSS (square root of sum of squares) of the individual moments due to the inertial and kinematic loads

b. For $T_b < T_g$: The peak bending moment in the pile can be computed by the algebraic addition of inertia and kinematic components.

6.6.4.2 Design approaches for pile foundation

The current design methods are based on pile design against bending failure due to lateral loads such as inertia and lateral spreading. Two main methods are used.

- Force-based method or limit equilibrium method: Lateral pressure acting on the pile is estimated and the response of the pile is evaluated in this method. Pile yielding and allowable deflection are also checked in this method.
- Displacement-based method, or p–y method, or seismic deformation method: Free-field ground displacement for an earthquake is evaluated up to pile depth. The obtained displacement profile is applied on the pile and its response is evaluated.

6.6.4.2.1 Force method, or limit equilibrium method

This method is recommended by specifications in [85] for the design of pile foundations in liquefying soils under lateral spread. In this method, the lateral forces are evaluated to obtain the pile response. The lateral forces are based on passive earth pressure in the non-liquefying region and 30% of the overburden pressure in the liquefying region. It is reported in [86] that the basis of these pressure distributions was on the back calculation of case histories of pile performance in the 1995 Kobe earthquake. A simplified limit equilibrium method for computing maximum bending moment in a pile was presented in [87] and is given by

$$M_{max} = (0.5 A_p H_p + A_c H_c) p_L \tag{6.13}$$

where:
A_p = area of pile exposed to lateral liquefied soil pressure
H_p = length of pile exposed to lateral liquefied soil pressure
A_c = area of pile cap exposed to lateral liquefied soil pressure
H_c = height of force F_c above the bottom of liquefied sand layer
F_c = lateral equivalent force on the pile cap
p_L = limiting liquefied soil pressure

6.6.4.2.2 p–y Method or seismic deformation method

This method is dependent on the concept of a Winkler beam on elastic foundation–type models, which require free-field displacement and are obtained from ground response analysis. The values of reduction in spring stiffness [89], which is based on factor of safety against liquefaction (F_L), are shown in Table 6.7.

The North American practice is to multiply p–y curves by a degradation factor p; this is called p-multiplier. This value usually ranges from 0.1 to 0.3. It decreases with pore pressure increase [89] and will become 0.1 when the excess pore water pressure is 100%. In [90], force-based analysis is compared with displacement-based analysis in the case of single piles subjected to lateral spreading problems. According to this study, the force-based method reasonably predicted the bending moments but underestimated the pile displacements. Also, the displacement method predicts both the pile bending moment and displacement reasonably accurately.

6.6.4.3 Analysis of pile in liquefying soil considering failure criteria

6.6.4.3.1 Bending criteria

The most commonly used model for predicting soil-pile response is by using the p–y curve technique recommended by [91], considering Tb > Tg, where kinematic interaction is predominant. The p-y curve required for solving the basic differential equation of laterally loaded pile is given by

$$EI\frac{d^4y}{dz^4} + E_s y = F \qquad (6.14)$$

where:

y = lateral deflection of pile
z = depth along the pile from top

Table 6.7 Reduction factors to stiffness degradation due to liquefaction

F_L range	Depth range (m)	D_E (Reduction Factor)
$F_L \leq 0.6$	$0 \leq x \leq 10$	0.00
	$10 \leq x \leq 20$	0.33
$0.6 \leq F_L \leq 0.8$	$0 \leq x \leq 10$	0.33
	$10 \leq x \leq 20$	0.67
$0.8 \leq F_L \leq 1$	$0 \leq x \leq 10$	0.67
	$10 \leq x \leq 20$	1.00

Source: JRA, Japanese Road Association, Specification for Highway Bridges, Part V, 1980.

EI = flexural rigidity of pile
E_s = soil modulus
F = applied force per unit length of pile

In earthquake engineering, this equation is modified by [92,77].

In case of liquefying soil, the subgrade soil modulus is degraded, and the degradation of k_{hn} with increasing displacement is expressed by [93,94] and is described by

$$EI \frac{d^4 y}{dz^4} = -pD$$

$$EI \frac{d^4 y}{dz^4} = -k_h (y - y_g) D \tag{6.15}$$

where:
y_g = ground displacement
D = diameter of pile
k_h = subgrade soil modulus

Variation of horizontal subgrade modulus, k_{hn} (for non-liquefied soils) with depth in the soil deposits is correlated with the SPT N values. The modulus of subgrade reaction for non-liquefied soils k_{hn} proposed by [92] and [85] is given by

$$k_h = k_{hn} S_f \tag{6.16}$$

where S_f = scaling factor of liquefied soil, and

$$k_{hn} = 80 E_0 B_0^{-0.75} \tag{6.17}$$

where:
E_0 = 0.7N
k_{hn} = modulus of subgrade reaction [MN/m²>]
E_0 = modulus of deformation [MN/m²]
N = SPT value
B_0 = diameter of the pile [cm]

When the soil liquefies, the stiffness of the soil degrades. The case studies indicate that the modulus of subgrade reaction for the laterally spreading soils can be reduced by use of a scaling factor, also termed as stiffness degradation parameter, S_f, which varies from 0.001 to 0.01 [95], as compared to normal soil conditions where there is no liquefaction.

The Winkler type model holds good for soil subjected to large displacement where lateral force from soil is proportional to the relative motion between pile and soil, respectively:

$$p = k_h \left(y - g(z,x) \right) \tag{6.18}$$

where:

g(z,x) = permanent ground displacement profile with depth z near the pile

x = the lateral distance from pile

In the case of lateral spreading near a waterfront structure, the permanent horizontal ground displacement generally decreases from waterfront toward inland. The distance of lateral spreading from the waterfront, Ls, is given by [93]

$$\frac{L_s}{L_2} = (25 \text{ to } 100) \frac{g_0}{L_2} \tag{6.19}$$

where:

L_s = effected distance of laterally spread ground from waterfront

L_2 = depth of liquefied soil layer [m]

g_0 is the permanent horizontal ground displacement at the waterfront and is described by

$$g_0 = \min(g_{max}, g_w) \tag{6.20}$$

where:

g_w = displacement of the quay wall

g_{max} is the maximum possible permanent ground surface displacement of the liquefied soil and is found out using the relation [96], which is given by

$$g_{max} = 0.75 \left(L_2 \right)^{0.5} \left(s_l \right)^{0.33} \tag{6.21}$$

where:

g_{max} = maximum possible permanent ground surface displacement of the liquefied soil [m]

L_2 = depth of liquefied soil layer [m]

s_l = slope of the base of the liquefied layer, or the gradient of the surface topography, whichever is the maximum

The horizontal ground displacement at a distance x from the waterfront g_x is expressed in a normalized form [97] and is given by Equation 6.22:

$$\frac{g_x}{g_0} = \left\{ 0.5\frac{5x}{Ls} + \left[1 - 0.5\frac{5x}{L_s} \right] \right\} \frac{g_{rs}}{g_0} \tag{6.22}$$

where:

g_{rs} = permanent horizontal displacement of the level ground far away from the waterfront (may be assumed to be zero)

The permanent horizontal ground displacement profile with depth z at a distance x of a laterally spreading deposit, $g(z, x)$ may be approximated as [94] (Equation 6.23).

for $z < L_1, g(z, x) = g_x$

for $z \geq L_1$ and $z \leq (L_1 + L_2), g(z, x) = g_x \cos\left(\frac{\pi(z - L_1)}{2L_2} \right)$

for $z \geq (L_1 + L_2), g(z, x) = 0$ $\tag{6.23}$

6.6.4.3.2 Bending and buckling criteria

If during an earthquake, $T_b < T_g$, then both kinematic and inertial interactions are predominant. In such cases, piles will be subjected to both axial and lateral load, thereby acting as beam-column member. The presence of both lateral and axial loads may cause degradation of lateral stiffness, and, as axial load approaches critical value, beam deflections due to loss of lateral stiffness. The large deflection may induce the state of plasticity in the beam that may lead to early failure of the beam. This type of interaction is termed as bending-buckling criteria.

The governing differential equation considering both inertial and kinematic loading was given in [98]

$$EI\frac{d^4y}{dz^4} + P\frac{d^2y}{dz^2} = -k_h D(y - y_g) \tag{6.24}$$

where:

P = vertical loading applied at the pile head
y_g = permanent ground displacement
D = diameter of pile
k_h = subgrade soil modulus

The variation of degradation of subgrade modulus can be determined by [93] and [94] as mentioned above.

6.6.5 Performance of pile foundations during recent earthquakes

Some of the case histories available on pile performance during recent earthquakes are the Showa Bridge collapse and Yachiyo Bridge failure during the 1964 Niigata earthquake [99], the LPG tank and Hanshin expressway pier failure [71] during the 1995 Kobe earthquake, and the Harbour Master building failure at Kandla Port during the 2001 Bhuj earthquake [2]. In the majority of cases, lateral spreading was reported.

During a seismic event, a structure may fail due to its structural inadequacy, a foundation failure, or a combination of both. In such failures, the soil supporting the foundation plays an important role. The behavior of foundations during earthquakes is often dictated by the response of its supporting soil due to the ground shaking. In general, there are two types of ground response that are damaging to the structures.

1. Soil can fail typically by liquefaction, as occurred in the 1995 Kobe earthquake.
2. Soil amplifies the ground motion as in the 1989 Loma Prieta earthquake.

Table 6.8 shows the case histories of the responses of various pile foundations during noted earthquakes.

6.7 SEISMIC ANALYSIS OF PILE GROUPS

Piles are most commonly used in groups. For analysis of piles in groups, the soil–pile interaction should be taken into account. It affects the response of piles in two ways:

1. Cross-interaction among piles (pile–pile or pile–soil–pile interaction)
2. Influencing the key parameters such as soil stiffness because piles are mostly used in soft soil, which is susceptible to stiffness degradation during seismic events.

The simplified procedure to analyze the pile group suggested by [75,102] involves the following steps:

1. Calculate the free-field ground response analysis and evaluate the ground displacement along the soil depth and magnitude of maximum ground acceleration.
2. Perform static analysis by applying ground displacement over the pile length, which is kinematic loading, and by applying lateral load at the

Table 6.8 Summary of case history on pile foundation in different past earthquakes

No	Case history	Earthquake event	Pile Material	Diameter (m)	Length (m)	Lateral spreading	Pile performance	Reference
1	10 storey-Hokuriku	1964 Niigata	RCC	0.4	12	Yes	Good	[70,99]
2	Showa bridge		Steel tublar	0.6	25	Yes	Poor	
3	Yachivo bridge		RCC	0.3	11	Yes	Poor	
4	NHK building		RCC	0.35	12	Yes	Poor	[70]
5	NFCH building		RCC hollow	0.35	9	Yes	Poor	
6	Gaiko Ware House		RCC	0.6	18	Yes	Poor	[99]
7	Landing bridge	1987 Edgecumbe	PSC	0.4	9	Yes	Good	[1]
8	4 storey fire house	1995 Kobe	PSC	0.4	30	Yes	Poor	[100]
9	14 storey building in American park		RCC	2.5	33	Yes	Good	
10	Hansin Expressway pier		RCC	1.5	41	Yes	Good	[71]
11	LPG tank 106, 107		RCC	0.6	16	Yes	Poor	
12	LPG tank 101		RCC	1.1	27	Yes	Good	
13	Kobe Shimin hospital		Steel tublar	0.66	30	Yes	Good	[101]
14	Elevated pert liner railway		RCC	0.6	12	Yes	Poor	
15	3 storey building, Fukae @ Kobe		PSC	0.4	20	Yes	Poor	[72]
16	Harbour Master's building, Kandla port	2001 Bhuj	RCC	0.4	30	Yes	Poor	[2]

Source: Modified after Madabhushi S.P.G. et al., Geotechnical aspects of the Bhuj Earthquake, in EEFIT Report on the Bhuj Earthquake, Institution of Structural Engineers, London, 2005.

Note: RCC: Reinforced cement concrete; PSC: Prestressed concrete

pile cap, which is the product of pile cap mass and maximum acceleration obtained through ground response analysis.

6.7.1 Failure mechanism of pile group

Pile group is the most common foundation used in structures founded on soft and liquefiable soil deposit. The following cases of failure of pile group are considered.

6.7.1.1 Formation of plastic hinge both at top and bottom of pile group

For a pile group that is rock-socketed and passing through liquefiable sand layer, the liquefiable layer may lose its capacity due to loss of stiffness during a strong seismic event. The pile may fail by the formation of a plastic hinge at the connection between pile and pile cap, and at the point of pile socketing. The pile cap may fail due to rotation. The maximum bending moment is obtained at the pile head in the case of a fixed pile and at the boundary between the liquefiable sand layer and the non-liquefiable layer. Figure 6.19 shows four hinge failure mechanisms in a pile group where piles fail due to the formation of plastic hinges at pile heads as well as pile tips. If piles are resting on a dense sand layer and not rock-socketed, then no plastic hinge will form at the pile tips. In this mechanism, that pile cap may suffer lateral displacement and small movement may cause huge stress in the superstructure. Figure 6.19 also shows three hinge failure mechanisms where piles will fail due to the formation of plastic hinges and one of the piles will fail due to buckling instability. This type of failure mechanism may cause severe rotation of pile cap, which leads to severe stress in the superstructure.

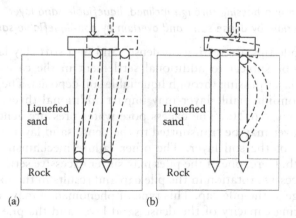

Figure 6.19 Pile group failure mechanism in (a) four hinge failure mechanism (b) three hinge failure mechanism.

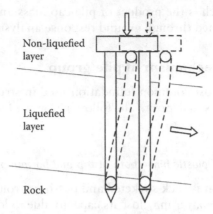

Figure 6.20 Pile group failure mechanism in laterally spreading ground with non-liquefiable layer as one of the mayor force.

6.7.1.2 Pile group passing through inclined, liquefiable sand layer underlain by bedrock and overlain by non-liquefiable sand

During an earthquake event, the liquefiable sand layer liquefies and tends to spread laterally and hence a large lateral load will be generated on the pile group due to passive earth pressure from a non-liquefiable layer lying over it. Thus, a large lateral load in addition to the axial load may lead to the bending of the pile, resulting in the formation of a plastic hinge, as shown in Figure 6.20. The maximum bending moment can be obtained at the point of connection between pile cap and pile and point of rock socketing.

6.7.1.3 Pile group passing through inclined, liquefiable sand layer underlain by dense sand and overlain by non-liquefiable sand

When the pile tips are resting on dense sand or a stiff clay layer, the pile group may be subject to additional settlement in the case of the piles' underlying layer passing through liquefiable soil deposits. The lateral loads from the non-liquefiable layer may impart additional thrust on piles, as shown in Figure 6.21a. The excess pore water pressure generated in the loose soil layer may be transmitted to the dense sand layer, which causes a softening of the soil layer. The other failure mechanism is shown in Figure 6.21b. Here, one of the piles may suffer excessive settlement, which leads to excessive rotation in the pile cap and results in the formation of a plastic hinge at the pile cap. This type of phenomenon may cause the loss of the bearing capacity of the dense sand layer and the pile may experience excessive settlement on one side that causes the formation of a plastic hinge in the pile.

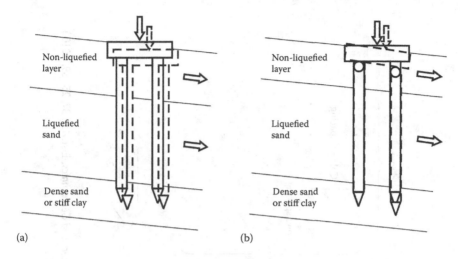

Figure 6.21 Failure mechanism in pile group (a) and bearing failure (b). Combination of local bearing failure and plastic mechanism.

6.7.2 Pile group pseudo-static analysis

Assumption of analysis:

1. Soil is assumed to be an ideal, isotropic, elastic material having Young's modulus (E_s) and Poisson's ratio (μ).
2. Pile is assumed to be a thin strip of width (b), length (L), and constant flexural rigidity ($E_p I_p$)
3. Each pile is divided into small elements of length (δ), except those at top and bottom element whose length is ($\delta/2$).
4. Normal stresses developed between the pile and soil are only considered in the analysis; shear stress is not.
5. The horizontal stress (p) developed in a pile is assumed to be constant across its length.
6. Soil at the back of a pile near the surface adheres to the pile and horizontal displacement between soil and pile is the same and equated at element centers, except for the top and bottom tip of pile.

The generalized differential equation 6.24 in finite difference form for bending thin beams for a pile is shown in Figure 6.22.

Soil displacement can be calculated based on [103], which gives the value of displacements within a semi-infinite, elastic, isotropic, homogeneous mass caused by horizontal point load. The soil displacement for all points along pile (x) in the group, which is induced by soil–pile and pile–soil–pile interaction, is given by Equation 6.25:

$$\frac{E_p \, I_p}{\delta^4}[D]\{u_p\} = -d\{p_p\} \tag{6.25}$$

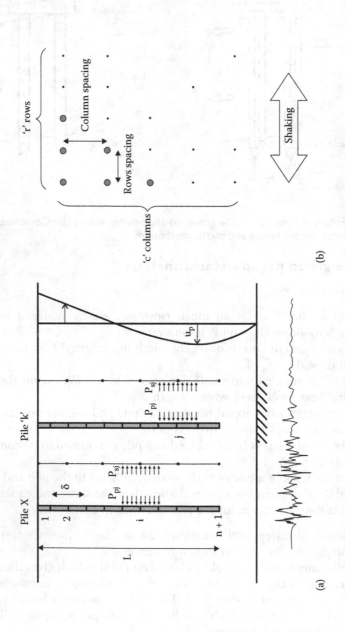

Figure 6.22 Pile group arrangement: (a) cross-section, (b) Plan view. (From Elahi et al., *Computers and Geotechnics*, Elsevier, 37, 25–39, 2010.)

where:
p_p = vector of pressure acting on pile
u_p = vector of pile displacement
D = matrix of finite difference coefficient
δ = equal element length

The displacement compatibility equation between pile and soil can be written by combining Equations 6.25 through 6.27:

$$\{u_s\}_x = \{u_e\}_x + [I_s]_{xx}\{p_s\}_x + \sum_{k=1\neq x}^{r\times c}[I_s]_{xk}\{p_s\}_k \tag{6.26}$$

where:
$\{u_s\}$ = vector of soil horizontal displacement
$\{u_e\}$ = vector of external soil movement
$\{p_s\}$ = vector of pressure acts on soil
$[I_s]$ = $(n+1)\times(n+1)$ matrix of soil displacement influence factors with its components, that is, $I_{s,ij}$,

$I_{s,ij}$ = displacement of element (i) due to applied horizontal unit force in the element J of pile, c, and r is number of row and column, respectively, in a group

$$\left[[U] + \frac{E_p I_p}{d\delta^4}[I_s]_{xx}[D]\right]\{U\}_x + \frac{E_p I_p}{d\delta^4}\sum_{k=1\neq x}^{r\times c}[I_s]_{xk}\{u\}_k = \{u_e\}_x \tag{6.27}$$

The equation can further be written as

$$\{U\}_x + \frac{E_p I_p}{d\delta^4}\sum_{k=1\neq x}^{r\times c}[I_s]_{xk}[D]\{u\}_k = \{u_e\}_x \tag{6.28}$$

Equation 6.27 gives an $(n+1)$ equation for $(n+1)$ displacements. For two end nodes, two auxiliary points are required, total unknown is now $(n+5)$, then for $(r\times c)$ total number of unknown equation and total number of unknown displacement is $r\times c\times(n+5)$. Boundary condition at the pile end is applied to solve these equations.

6.8 SEISMIC SOIL–PILE STRUCTURE INTERACTION

Seismic soil–pile structure interaction refers to the effect of superstructure-supporting pile foundations on the motion of the superstructure. Figure 6.23 shows different cases of failure of pile foundation under seismic events. The seismic response of pile foundations is very complex and involves the interaction between superstructure and pile foundations (inertial interaction),

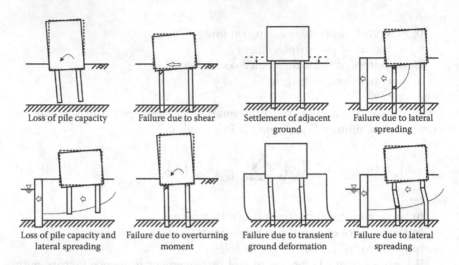

Loss of pile capacity   Failure due to shear   Settlement of adjacent   Failure due to lateral
                                                        ground   spreading

Loss of pile capacity and   Failure due to overturning   Failure due to transient   Failure due to lateral
lateral spreading   moment   ground deformation   spreading

Figure 6.23 Soil–pile–structure interaction under seismic events. (From Tokimatsu et al., *Proceedings on Geotechnical Earthquake Engineering and Soil Dynamics*, pp. 1175–1186, 1996.)

interaction between piles and soil (kinematic interaction), and seismically induced pore water pressures and nonlinear responses of soil to strong seismic motion. Many researchers have made contributions in this field [104–110]. Different approaches are available to incorporate dynamic soil–pile interaction, but they are usually based on the simplified assumptions that the soil is linear elastic or viscoelastic and the soil is perfectly bonded to a pile. The bonding between the pile and soil is rarely perfect in the actual filed condition. Furthermore, the soil mass immediately adjacent to the pile can undergo a large degree of deformation, which would cause the soil-pile system to behave in a nonlinear manner. A lot of efforts have been made in the recent past to model the soil–pile interaction by using the 3D finite-element method (FEM) and the three-dimensional finite difference method (FDM). However, it is too complex to model a pile group in nonlinear soil.

6.8.1 Three methods of analyzing seismic soil–pile structure interaction

6.8.1.1 Elastic continuum method

The elastic continuum method, which is based on Mindlin's solutions for point load to semi-infinite domain, was first used to analyze the soil–pile interaction problem [114]. This approach has been modified by various researchers considering the effect of superstructure, soil stiffness degradation, soil non-homogeneity, material damping, and so on.

6.8.1.2 Nonlinear Winkler foundation method

A linear elastic beam-column representing the pile attached to fully non-linear p-y springs and dashpots representing the surrounding soil are the basic assumptions of this model. An uncoupled approach to performing the dynamic analysis of offshore structures was devised by [115]. Similar studies have been made into the dynamic response of soil–pile structure. The most simplified approach is to obtain free-field acceleration-time history by using free-field site response analysis, then corresponding displacement time histories are applied to the nonlinear p–y springs to obtain the dynamic responses of the foundation. The nonlinear Winkler foundation model is effective and popular in design codes due to its simplicity; however, this method ignores three-dimensional interaction effects of soil-pile contact by using a two-dimensional simulation.

6.8.1.3 Finite element method

The FEM is capable of performing either two-dimensional or three-dimensional fully coupled analyses of pile foundations, which can simulate any arrangement of soil, pile, and superstructure. The FEM includes two basic methods: substructure method and direct method. Substructure method includes structure, pile foundation, and nonlinear soil element. The infinite soil is modeled as a regular, layered, homogeneous, semi-infinite boundary and is considered by a rigorous interaction through force-displacement relationship. Integration of the force-displacement relationship of the unbounded domain into equations of motion of the structure gives the dynamic analysis of soil–pile structure systems. Direct method includes the finite element region, which contains the structure, pile foundations, and soil profile up to the artificial boundary. The semi-infinite half-space, soil, is represented by artificial boundary conditions, simulating the wave propagation and energy dissipation so that no wave reflection exists from the outwardly propagating waves.

6.8.2 Soil–pile structure interaction approach described by various researchers

Stepwise soil–pile structure interaction in liquefiable soil is described in [93], as shown in Figure 6.23:

1. Before the development of pore water pressure, the inertia force from the superstructure may dominate. This was referred to as step I.
2. Kinematic forces from the liquefied soil start acting with increasing pore pressure. This was referred to as step II.
3. Toward the end of shaking, kinematic forces would dominate and have a significant effect on pile performance, particularly when permanent displacements occur in laterally spreading soil.

6.8.2.1 Concept of pile failure by [71]

Ref. [71] had summarized the seismically induced loading on the pile by introducing the concepts of top-down effect and bottom-up effect. At the onset of shaking, the inertia forces of superstructure are transferred to the top of the pile, and ultimately to the soil. It is assumed in [71] that during the main shaking, sandy soils had not softened significantly due to lique-faction and that the relative movement between the piles and ground were small. However, the author postulated that if ground motion was suffi-ciently high such that the induced bending moment in the piles exceeds the limiting value, the piles might fail. Since the load comes from the inertia force of superstructure, it was referred to as a top-down effect. He con-cluded that the observed failure of a pile in the upper portion after an earthquake might be attributed to this effect. The author also reported that the onset of liquefaction took place approximately at the same time that peak acceleration occurred in the course of seismic load application having an irregular time history. Thus, in sloping ground, the softened ground will start to move horizontally following the onset of liquefac-tion. Under this condition, lateral forces would be applied to the pile body embedded in the ground, leading to deformation of the pile in the direc-tion of the slope. The author assumed that seismic motion had already passed the peak and that shaking may have still been persistent with lesser intensity and therefore that the inertia force transmitted from the super-structure would be insignificant. Under such a loading condition, the max-imum bending moment induced by the pile may not occur near the pile head but at a lower portion, at some depth, and this is referred to as the bottom-up effect.

6.9 SEISMIC ANALYSIS OF COMBINED PILE-RAFT FOUNDATION (CPRF)

Seismic analysis of a CPRF is very complex and needs proper understand-ing of pile–soil–raft interaction. The lateral load induced on a CPRF during earthquake motion is jointly shared by raft and pile component. Hence, the stiffness parameter and connection condition between pile and raft has to be defined carefully by performing sensitivity analysis. Figure 6.24 shows the generalized CPRF model subjected to seismic excitation.

It is observed by various researchers that horizontal stiffness of a CPRF is larger than that of a pile group with the same configuration as a CPRF, because a raft acts efficiently as a horizontal displacement reducer. It is also noted that the bending moment of piles in a CPRF is less than that of piles in a pile group. The rotation of a CPRF increases as the rigidity of connection condition increases. The horizontal load carried by a CPRF does not have significant influence on the pile head connection, whereas it

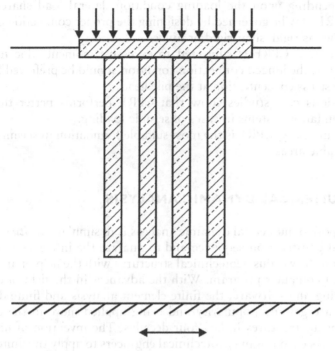

Figure 6.24 Generalized pile-raft model subjected to seismic excitation.

has significant influence on the horizontal load proportion, where the raft shares a larger proportion of the load as the connection condition rigidity increases.

For the design of a piled raft foundation subjected to lateral loading where average settlement is permitted with limited difference settlement, interaction between building and foundation component is employed. The seismic analysis and design of the CPRF is done by using the results obtained by centrifuge modeling, numerical simulations, and case studies.

6.9.1 Advantages of CPRF under dynamic conditions

- Improvement of serviceability of shallow foundations by reducing settlement, differential settlements, and tilt, and avoiding eccentricities and making loads centric, by concentrating piles around highly loaded area.
- Lateral load induced during dynamic condition will be shared by raft and pile, depending on their stiffness.
- If a pile passes through liquefiable soil deposit, raft provides buckling stability to pile foundation in case pile is subjected to excessive axial load.

- Depending upon the loading condition, lateral load shared by the CPRF may be governed by designing the proper connection condition either as rigid, semi-rigid or hinged.
- In case a CPRF is subjected to excessive moments due to seismic events, the hinged connection condition should be preferred to reduce the stress concentration at the pile head.
- Various case studies show that CPRF performs better than other foundation systems in cases of seismic loading.
- In summary, CPRF is the most suitable foundation in seismically vulnerable areas.

6.10 NUMERICAL DYNAMIC ANALYSIS

The purpose of numerical dynamic analysis is to simulate *in situ* soil conditions and construction sequence and to analyze the influence of dynamic loading on the various geotechnical structures with the help of an available numerical computer program. With the advances in the field of computer engineering and software, the finite element analysis and finite difference analysis are gaining popularity and being applied more in the design of geotechnical structures in last four decades. The invention of high-speed computers is encouraging geotechnical engineers to apply the finite element analysis for solving complicated problems including various dynamics. Different researchers have used different approaches for dynamic analyses of pile, pile group, and pile raft foundations, which include the application of seismically induced lateral loads to the top of the pile head, or applying the acceleration, force, or displacement time history at the base of a soil model to capture the response of such structures. Two different types of approach are used to analyze deep foundation problems.

6.10.1 Steps to be followed for the design of single pile, pile group and CPRF

- Soil investigation data has to be properly checked before selecting any soil parameters and constitutive soil model to simulate realistic soil behavior.
- If stiffness parameters of soil are constant values, which is normally for the case of sand, then the Mohr-Coulomb model is the preferred choice, but if stiffness parameters of soil are depth dependent or the soil is soft clay, care has to be taken in choosing the constitutive model to simulate actual field condition. The hardening soil model and the soft clay constitutive model can be chosen.
- To model the liquefaction condition, Finn and Byrne models can be chosen, which are dependent upon the relative density and penetration resistance of soil.

- The size of the numerical model should be greater than about 2.5 times the length of longest pile for a static case and should be increased further for lateral loading and seismic loading cases, based on engineering judgment, to eliminate the boundary effects.
- Pile spacing, pile diameter, and pile position may be decided on the basis of procedure and parametric variation. Feasibility of CPRF must be checked based on settlement and differential settlement responses observed and compared with the conventional raft and pile foundation responses.
- For structural components, that is, raft and piles, high stiffness value has to be chosen as per the characteristic strength of raft and pile material. To obtain structural response, pile and raft should be modeled as structural elements, respectively.
- Consideration of uniformly distributed loads may be useful for preliminary design and proportioning of foundation components.
- Seismically induced loads can be replaced by equivalent static horizontal load, which is equal to the seismic acceleration coefficient times the maximum superstructural load acting on to the foundation unit. In another way, the real acceleration-time history may also be applied at the base of the soil model and the response of the CPRF can be observed. The acceleration particular to the site can be obtained by performing a ground response analysis from the results of a seismic hazard analysis. For further details about seismic hazard analysis, see [111–113].
- Suitable CPRF code like ISSMGE combined pile-raft foundation guidelines (2013) has to be followed properly for static design and can be updated further for seismic loading conditions.

6.10.2 Numerical dynamic analysis of oil tank foundation: A case study

A seismic design for pile foundations to support an oil tank in a seismically active region of Iraq was performed [117]. Geotechnical investigations revealed 2.5 m of medium dense sand layer followed by a layer of soft to stiff clay having a plastic limit more than 20% up to 24 m, and a dense sand layer extending to greater depth. The typical soil property is shown in Figure 6.25. An oil tank of height 15 m and diameter 23.5 m was proposed for the site. The dead loads of the oil tank, empty and full, under operating conditions were 1400kN and 62000kN, respectively. The characteristic strength of concrete was chosen as 35 N/mm². A finite element program, PLAXIS3D was chosen to simulate *in situ* conditions. Soil was modeled on the Mohr-Coulomb model, pile and pile cap were modeled as embedded pile element and plate element, respectively, as shown in Figure 6.26. A total of 89 piles of diameter 800 mm and length 26 m were used below the tank. The numerical model was first validated with a test conducted at the

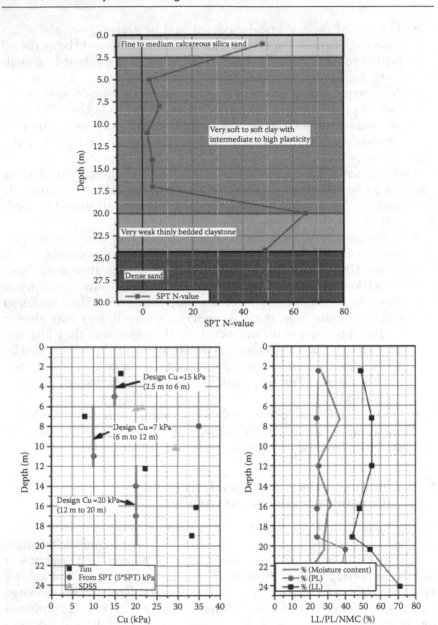

Figure 6.25 Geotechnical properties along soil depth. (From Kumar et al., *Disaster Advances*, 8(6), 33–42, 2015.)

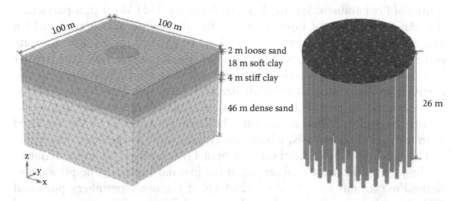

Figure 6.26 3D view of numerical model of pile foundation for oil tank in PLAXIS3D. (From Kumar et al., *Disaster Advances*, 8(6), 33–42, 2015.)

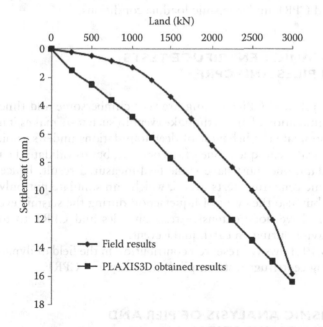

Figure 6.27 Comparison of load–settlement curve obtained in PLAXIS3D and in the field. (From Kumar et al., *Disaster Advances*, 8(6), 33–42, 2015.)

site. Figure 6.27 shows the load–settlement curve obtained in the field test, and the numerical simulation shows a reasonable validation of the numerical model. Axial load in pile varies from 510kN to 697kN from center pile to periphery pile. The total differential settlement of 3 mm was observed under static loading conditions.

Seismic analysis was carried out after completion of the static analysis by applying synthetically generated earthquake input motion based on the results of Probabilistic Seismic Hazard Analysis (PSHA) for that particular site. An earthquake of input motion of peak ground acceleration 0.15 g and 54 s duration was applied at the base of the soil model. For accurate wave transmission through the soil model, the mesh size was kept smaller than one-tenth of the wavelength associated with the highest frequency of earthquake input wave. The mesh size of 1.92 m was kept by considering above criteria. A viscous boundary that contains dampers was adopted to absorb the multiple wave reflection. The Rayleigh damping of 5% was used to in the modeling process, which was typical value for geologic material.

The maximum differential settlement of 4 mm is observed under dynamic loading. Amplification is observed at the ground surface. The piles experienced maximum axial load of 1284 kN at the outer periphery piles and 772 kN at the central piles, which dominate the design. It was also observed that differential settlement was within permissible limits of 1/300.

Table 6.9 shows the numerical study carried out to analyze pile, pile group, and CPRF under seismic loading conditions.

6.11 DYNAMIC CENTRIFUGE TESTS ON PILES AND CPRF

Full-scale pile and CPRF testing are very cumbersome and time-consuming, and simulation of an earthquake event in such tests makes it more complex. To simulate the behavior of deep foundations under seismic loading, the numerical technique is one of the options, but to validate the numerical analysis data, one must have some field-measured result, hence the need for dynamic centrifuge tests arises, which can simulate not only the state of stress, but also the event of liquefaction during the seismic event. Many researchers have done extensive tests on piles and CPRF to understand their behavior during an earthquake event.

Table 6.10 shows the research contribution in the field of dynamic analysis by using centrifuge tests on pile, pile group and CPRF.

6.12 SEISMIC ANALYSIS OF PIER AND WELL FOUNDATION

Well foundation (also known as Caisson foundation) is generally used for the deep foundations of railway and road bridges over rivers, aqueducts, and other structures. This foundation system is often when scouring of a riverbed is a major concern. The geometric and flexural properties make such foundations suitable over pile foundations. They are also able to carry large lateral and moment loads. Seismic response of this foundation type depends upon the *in situ* soil characteristics and spatial variations of earthquake motion at different depth of soil, and nonlinearity at soil-well

Table 6.9 Numerical study carried out for analyzing the behavior of pile, pile group and CPRF under seismic loading conditions

Piles

Sr. No.	Ref.	Methodology	Results and description
1	[119]	Modeled a single pile for both fixed headed and free headed piles by using FLAC3D, carried out dynamic analysis of single pile in liquefied and non-liquefied soil deposits	Bending moment and lateral deflection in pile Is more in liquefied soil deposits as compared to non-liquefied soil deposit, lateral deflection is more in case of free head pile as compared to fixed headed pile
2	[120]	Analyzed the response of vertical pile embeddec in dry dense sand subjected to cyclic lateral load by using centrifuge test and finite element program ABAQUS	The overall response of pile–soil system indicated the suitability of the chosen model for lateral response prediction; 1 × 2 pile group under cyclic lateral loading was capable of representing shadow effect of pile group
3	[121]	Conducted parametric study on single piles and pile groups embedded In a two layer sub-soil profile for evaluating the kinematic bending moment developed during earthquake by using finite element program VERSAT-P3D	Bending moment evaluated for both single pile and pile group with dynamic analyses in time domain under different sub-soil condition were more reliable compared to other simplified approaches
4	[122]	Extended the pseudo-static analysis of 2 × 2 pile group for accounting the instability effects of pile groups, carrying substantial axial load in liquefied soil of low stiffness using Riks post- buckling analyses method and implemented the same ABAQUS.	Flexible concrete piles were vulnerable to significant amplification and ultimate collapse at lateral loads; for stiffer and stronger steel piles, amplification was found to be low at maximum working load
5	[123]	Analyzed the influence on the lateral response of piles istalled in sandy soil by using 3D-GEOFEM; pile was modeled as linear elastic material and soil was modeled by using Drucker-Prager model	Vertical load had marginal influence on lateral response of piles in loose sand; Percentage improvement in lateral load capacity for measuring the influence of vertical load on lateral response of piles in loose and dense sand
6	[124]	Studied the effect of nonlinear behavior of soil and separation at soil-pile interface of single pile and pile group by using FORTRON code (3DnPILE), seismic analysis were conducted by applying harmonic load and transient excitation to pile	Nonlinearity suppressed the wave interference effect among the piles in group and significantly reduced the stiffness at excitation frequencies; the effect of soil nonlinearity depended upon the frequency of

CPRF

1	[125]	Analyzed pile group and CPRF under sinusoidal loading and El-Centro earthquake time history by using ABAQUS finite element program	Pile groups experience more settlement, lateral displacement, deformation and bending moment as compared to CPRF under both sinusoidal and earthquake loading.

Table 6.10 Research contribution in the field of dynamic analysis

Piles			
Sr. No.	Ref.	Methodology	Tests details	Results
1	[126]	Performed dynamic centrifuge test for 2×2 pile groups in three layered laterally spreading soil profile consisting of non-liquefiable cohesive crust overlying loose liquefiable sand, overlying dense sand	Slope boundary conditions and presence of axial loads were given considerations	Axial loads reduced the lateral capacity of piles to resist kinematic lateral loads from spreading soil, since they induced an additional P-Δ moment in pile; liquefaction causes excessive settlement of pile groups which leads to failure.
2	[127]	Conducted dynamic centrifuge test for investigating seismic behaviour of building having semi-rigid pile head connection	Test conducted at 50 g and Rinaki wave used having maximum acceleration of 200 cm/s² in the prototype scale	Semi-rigid pile head connection reduces the bending moment of piles before and after liquefaction, maximum acceleration in semi-rigid pile head connection was smaller when compared with fixed head connection
3	[114]	Performed dynamic centrifuge testing for investigating pile stability by examining the behaviour of pile groups subjected to lateral loads	Tests conducted at 80 g, model pile groups considered had two degree of freedom for lateral movement, subjected to dynamic load by using stored angular momentum actuator	Pile–soil flexibility had a strong influence on amplification occurring during test and high axial loads or amplification lead to unstable collapse of pile groups, developed of excess pore pressure altered the shape of supporting soil
4	[128]	Performed series of centrifuge test on four-pile group for free-headed, thin flexible cap, thick cap and monolithic cap	Test performed at 50 g	For free headed pile, zero bending moment at pile head with maximum bending moment near the mid height, negative bending moment observed at the pile head in

5	[129]	Eight dynamic model test on 9 m radius on pile groups (2×1 and 2×3) in and centrifuge and pile liquefiable and lateral spreading ground	Pile diameter ranges from 0.36 m to 1.45 m for single pile and 0.73 m to 1.17 m for pile groups; tested at peak base acceleration of 0.13 g t 0.1 g.	In all cases, the maximum bending moment is obtained at the boundary between liquefied and non-liquefied soil layer.
6	[130]	Eight centrifuge model of vertical single piles and pile group subjected to earthquake induced liquefaction and lateral spreading.	Tests conducted at 50 g in 40% dense Nevada sand layer; prototype pile diameter=0.6 m; 40 cycles of uniform accelerations applied with prototype amplitude of 0.3 g and frequency of 2 Hz	Permanent lateral displacement=80 cm Maximum bending moment in pile and pile group was observed at the interface between liquefiable and non-liquefiable sand layer, bending moment increased with the presence of pile cap

Combined pile-raft foundation (CPRF)

1	[131]	Seismic response of piled raft system with flexible and stiff pile embedded in Kaolin clay	Tested at 50 g, four piles connected to rigid raft and superstructure load having load of 368ton, 605ton, 863ton; Three scaled earthquake having acceleration of 0.22 g, 0.052 g and 0.13 g	For both flexible and rigid pile-raft system, soft clay acted as inertial loading rather than providing support; for rigid pile bending moment transition from positive on the top and negative on the bottom; for flexible pile, below certain length negligible bending moment response observed
2	[132]	Carried out series shaking table tests by using geotechnical centrifuge to study the effect of rigidity of connection condition of pile raft and free standing pile-group	Tests conducted at 50 g and in Toyoura sand layer with 60-63% density. Model subjected to sinusoidal wave of amplitude 1 m/s² with frequency of 50 Hz for time period of 0.6 s.	Vertical load carried by raft was 60% and 55% of total load in case of rigid and hinged connection model; 20% more acceleration observed in case of rigid connection model and same response was observed for settlement also.

interface. Such foundations also suffer moderate to severe displacement in the event of liquefaction during an earthquake.

Well foundation is generally analyzed by modeling as simple linear Winkler springs by neglecting soil and interface nonlinearities. In practice, well foundation is modeled as two-dimensional plain strain or one-dimensional spring dashpot model. The one-dimensional approach is frequently used in practice due to its simplicity and easy implementation. The one-dimensional spring dashpot model, which incorporates soil and interface nonlinearity, is proposed in [133].

6.12.1 One-dimensional (1D) spring dashpot analysis of soil-well-pier foundation

The model considers Novak's and Veletsos's model as the basic model, as it is widely used for deep foundations. Figure 6.28 shows the four springs distributed along the length of a well foundation viz., translational and rotational spring dashpot, concentrated base translational, and rotational spring dashpot. Ref. [134] proposed the dynamic stiffness of unit length for an infinitely long rigid cylinder embedded in homogenous soil and subjected to translational and rotational modes of vibrations:

$$k_x = G\left[S_{x1}\left(a_0, \upsilon, D\right) + iS_{x2}\left(a_0, \upsilon, D\right)\right] \tag{6.29}$$

$$k_\theta = Gr^2\left[S_{\theta 1}\left(a_0, D_s\right) + iS_{\theta 2}\left(a_0, D\right)\right]$$

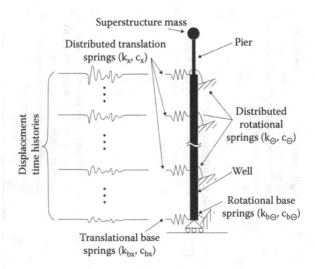

Figure 6.28 Mass-spring-dashpot model for soil-well-pier foundation. (From Mondal et al., *Earthquake Spectra*, 28(3), 1117–1145, 2012.)

where:

k_x, k_θ = the dynamic impedance of soil for translation and rotational vibrations, respectively

r_o = the radius of circular cylinder or equivalent radius

$a_0 = r_o\omega/v_s$ = dimensionless frequency

Vs = shear wave velocity of soil mass

υ = the poisons ratio of soil

D = material damping

Spring and dashpots were connected in parallel and distributed throughout the embedment. The base resistance is modeled by introducing a spring at the base, and the stiffness of calculating base resistances can be obtained from the equation proposed by [136,137]

$$k_{bx} = \frac{8Gr_o}{2-\upsilon} \qquad (6.30)$$

$$c_{bx} = \frac{8Gr_o}{2-\upsilon} \cdot \frac{0.6a_0}{\omega}$$

$$k_{b\theta} = \frac{8Gr_0^3}{3(1-\upsilon)} \cdot K'_{b\theta}$$

$$c_{b\theta} = \frac{8Gr_0^3}{3(1-\upsilon)} \cdot \frac{0.35a_0^3}{\omega(1+a_0^2)}$$

$$K'_{b\theta} = \begin{cases} 1-0.2a_0 & \text{for } a_0 \leq 2.5 \\ 0.5a_0 & \text{for } a_0 > 2.5 \end{cases}$$

The system is subjected to displacement history corresponding to the soil layer estimated from equivalent linear analysis by using SHAKE2000, and the responses of the well foundation can be obtained.

6.12.2 Finite element analysis of soil-well-pier foundation

Soil-well interface behavior during ground shaking was studied in [135], which also evaluated the significance of interface nonlinearity by using the two-dimensional finite element model considering soil and interface nonlinearity under both full and partial embedment of well foundation. Soil was assumed to be cohesionless and tested for both dry and saturated condition. Figure 6.29 shows the two-dimensional geometry of the model used in the study. The height of the pier was 13.47 m, and the depth of the well was 50 m. For a full embedment condition, a 20 m thick layer of medium

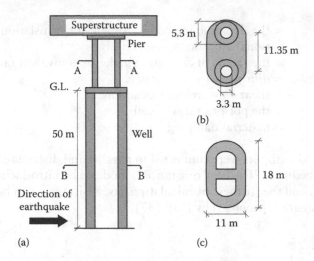

Figure 6.29 Geometry of well foundation and Piers used for the present study. (a) Elevation; (b) section A-A; (c) section B-B. (From Mondal et al., *Earthquake Spectra*, 28(3), 1117–1145, 2012.)

sand followed by a 30 m thick layer of medium dense sand followed by a 50 m thick layer of medium dense sand and bedrock extending to greater depth is considered. For partial embedment, soil was assumed to be eroded from the top. Figure 6.30 shows axisymmetric modeling details of the soil-well-pier foundation. Validation of the model was carried out with the 1940 El-Centro earthquake motion. After successful validation, the same model was further analyzed under peak ground acceleration of 0.2 g, 0.4 g and 0.6 g, respectively.

Results show that depth of separation increased with increases in embedment ratio. Maximum sliding was observed near the ground surface. The interface nonlinearity does not show any significant influence on the design displacement and resultant forces in the well and piers.

6.13 CODAL PROVISIONS

6.13.1 Codal provision for ground response analysis

Ref. [138] proposed very broad classification of soil sites to consider local site effects. The response spectra for three different types of soil, namely, rock and hard soil, medium soil and soft soil at founding level have been specified. There is no definite criterion for seismic classification of soil. International codes [139] adopt site classification based on the average shear wave velocity in the upper 30 m of soil deposit (Vs30). These codes have proposed site coefficients based on the intensity level of bedrock motion

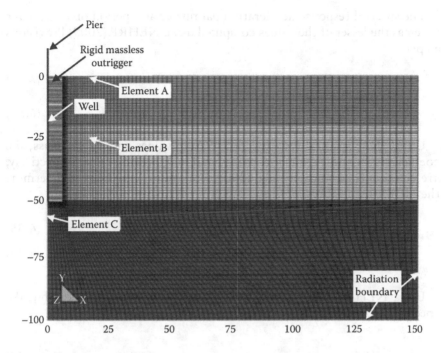

Figure 6.30 Two-dimensional view of soil-well-pier foundation in SAP 2000. (From Mondal et al., *Earthquake Spectra*, 28(3), 1117–1145, 2012.)

to compute surface level seismic hazard parameters. Hence, in order to capture local site effects, site-specific seismic hazard analysis and comprehensive ground response analysis is imperative to quantify seismic hazard parameters.

6.13.1.1 NEHRP (2009)

The NEHRP (2009) provisions provide the uniform hazard ground motion (SSUH and S1UH) maps, risk coefficient (CRS and CR1) maps, and deterministic ground motion (SSD and S1D). The subscripts "UH" and "D" indicate uniform hazard and deterministic values of the spectral response parameter at short period (0.2 s) and long period (1.0 s), Ss and S1. The CRS and CR1 are the mapped risk coefficients at short and at long periods.

The spectral response acceleration parameter at short periods, SS, is taken as the lesser of the values computed using NEHRP (2009) Provisions as per

$$S_S = C_{RS} \, S_{SUH} \tag{6.31}$$

$$S_S = S_{SD} \tag{6.32}$$

The spectral response acceleration parameter at a period of 1.0 s, S1, is taken as the lesser of the values computed using NEHRP (2009) Provisions as per

$$S_1 = C_{R1}\, S_{1UH} \tag{6.33}$$

$$S_1 = S_{1D} \tag{6.34}$$

Using, these spectral response acceleration values and the site class, site coefficients Fa and Fv, at short period and long period (1.0 s), respectively, are determined. Using NEHRP (2009) Provisions, equations to determine the MCER spectral response acceleration parameters are

$$S_{MS} = F_a S_s \tag{6.35}$$

$$S_{M1} = F_v S_1 \tag{6.36}$$

Using NEHRP (2009) Provisions to determine the design earthquake spectral response, the acceleration parameters are given as

$$S_{SD} = \frac{2}{3} S_{MS} \tag{6.37}$$

$$S_{D1} = \frac{2}{3} S_{M1} \tag{6.38}$$

The NEHRP (1997) Provisions are part of IBC (2003). The NEHRP (2009) Provisions are a major technical modification to ASCE 7 (2005). The NEHRP 1997, 2000, and 2003 Provisions adopted a "Uniform Hazard" approach as against the "Risk-Targeted" approach adopted in the NEHRP (2009) provisions to determine the Maximum Considered Earthquake hazard parameters. In the "Risk-Targeted" approach, the ground motion parameters are adjusted to provide a uniform 1% risk of collapse in a 50 year period for a generic building, as opposed to a uniform return period for the seismic hazard (FEMA P-751 2012).

6.13.1.2 ASCE 7 (2010) [140]

The ASCE 7 (2010) standard addresses the key issues of the seismic design of buildings and other structures by characterizing the earthquake ground motion. The effects of earthquake motions are resisted by designing and constructing the structures as per the standards recommendation with respect to the seismic ground motion values, design response spectrum, importance factor and risk category, seismic design category, and geologic hazard and geological investigation. The ASCE 7 (2010) standard criteria for seismic ground motion values are described below.

The acceleration parameters SS and S1 are determined from the 0.2 s and 1.0 s spectral response accelerations maps provided in ASCE 7 (2010) and from the USGS website http://earthquake.usgs.gov/designmaps. Where, SS = mapped MCER spectral response acceleration parameter at short periods, S1 = mapped MCER spectral response acceleration parameter at a period of 1.0 s. The MCER is the risk-targeted maximum considered earthquake ground motion response acceleration that represents the most severe earthquake effects that result in the largest maximum response to horizontal ground motions and with adjustment for targeted risk.

The MCER spectral response acceleration parameters for short periods (SMS) and at 1.0 s (SM1), adjusted for site class effects, are determined by

$$S_{MS} = F_a S_s \tag{6.39}$$

$$S_{M1} = F_v S_1 \tag{6.40}$$

where Fa and Fv are the site coefficients at a short period (0.2 s) and a long period (1.0 s), respectively.

The design earthquake spectral response acceleration parameters at short period, SDS, and at 1.0s period, SD1, are determined by

$$S_{SD} = \frac{2}{3} S_{MS} \tag{6.41}$$

$$S_{D1} = \frac{2}{3} S_{M1} \tag{6.42}$$

ASCE 7 (2010) also outlined the site-specific risk-targeted maximum considered earthquake (MCER) ground motion procedure that calculates MCER spectral response acceleration at any period that can be taken as the lesser of the spectral response accelerations from the probabilistic and deterministic ground motions. The probabilistic and deterministic ground motions considered in ASCE 7 (2010) are as follows.

- The probabilistic ground motions shall be taken as risk-targeted ground motions, in terms of the uniform hazard (2% in 50 year) ground motions.
- The deterministic ground motions are 84th percentile ground motions.
- The probabilistic and deterministic ground motions are redefined as maximum-direction ground motions, in lieu of geometric mean ground motions.

The design response spectrum and design acceleration parameters are then evaluated accordingly. The site-specific ground motion hazard-analysis procedure considering the MCER ground motions can be found in Chapter 21 of ASCE 7 (2010) standard.

6.13.1.3 Indian standard code (IS 1893-Part I, 2002)

The seismic effects due to earthquake ground motion can be addressed in the design of a building by computing the design seismic base shear (VB) with

$$V_B = A_h W \tag{6.43}$$

where A_h= design horizontal seismic coefficient for a structure as obtained by the expression

$$A_h = \frac{ZIS_a}{2Rg} \tag{6.44}$$

in which "Z" is a zone factor for the maximum considered earthquake (MCE). The zone factor is converted to zone factor for design basis earthquake (DBE) by using a factor 2 in the denominator of Z. "I" is the importance factor based on the functional use of the structure. "R" is the response reduction factor that depends on seismic damage performance and Sa/g is spectral acceleration coefficient for rock and soil sites based on natural period and damping of the structure. The design acceleration spectrum for vertical motion is taken as two-thirds of the design horizontal acceleration spectrum. The site classes specified as medium soil and soft soil have no definite criteria for distinction.

Table 6.11 describes the site class as per various seismic codes.

6.13.2 Design of pile foundation

Very few codes are available for designing pile foundations in liquefying soils. There are a few codes, such as the Japanese Highway Code of Practice [85,39] and Eurocode 8 (1998), which provide guidelines for analysis of pile foundations passing through liquefying soil deposits. The salient features of these codes are discussed here.

6.13.2.1 Development of Japanese code of practice (1972–1996)

Several bridges, such as the Showa and the Yachiyo bridges were damaged during the 1964 Niigata earthquake due to soil liquefaction. Based on experience and observation, the Seismic coefficient method was introduced in the Highway Bridge Specification [141] to take into account the effects of liquefaction [142]. The code was subsequently amended in 1980 and,

Table 6.11 Site classes as per various seismic codes

	Category	Description	Mean shear wave velocity for top 30 m $(V_{s.30 m})$
NEHRP (20D9) provisions	A	Hard Rock Soils	> 1500 m/s
	B	Firm to Hard Rock	760–1500 m/s
	C	Dense Soil, soft Rock	360–760 m/s
	D	Stiff soil	180–360 m/s
	E	Soft days	< 180 m/s
IS 1893 (Part I); 2002	Type I	Rock or Hard Soil: Well graded gravel and sand gravel mixtures with or without clay binders and clayey sands poorly graded or sand clay mixtures (GB, CW, SB, SW and SC as per IS 1498) having SPT .N" above 30.	
	Type II	Medium Soils: all soils with N between 10 and 30 and poorly graded sands or gravelly sands with little or no fines (SP as per IS 1496} with N > 15.	
	Type III	Soft Soils: All soils other than SP (as per IS 1498) with N < 10.	
IBC (2009)	A	A Hard Rock: Eastern United States sites only −V > 1500 m/s	
	B	Rock - 760 <V,≤ 1500 m/s	
	C	Very dense Soil and soft rock (undrained shear strength S_v > 100 kPa), 360 < V, ≤ 760 m/s	
	D	Stiff Soils with undrained shear strength 50 kPa < S, < 100 kPa or 15 < N < 50, 180 ≤ V, ≤ 360 m/s	
	E	Soft Soils, Profile with more than 3 m of soft clay defined as soil with PI > 20, moisture content w > 40%, undrained shear strength S, < 50 kPa and N < 15, V, < 180 m/s	
	F	Soils requiring Site specific: evaluations. Soils vulnerable to potential failure or collapse under seismic loading eg. Liquefiable soils, quick and highly sensitive clays, collapsible weakly cemented soils. Peats and highly organic clays 3 m or thicker layer. Very high plasticity clays 8 m or thicker layer with PI > 75. Very thick soft/medium stiff clays: 36 m or thicker layer.	

as a result, an alternative approach known as the "seismic deformation method" evolved. Following the damage of piled bridges in the aftermath of the 1995 Kobe earthquake, the Highway Bridge Specification was fully revised [143]. A new approach, with checks on lateral spreading, was introduced in this edition.

6.13.2.2 Japanese highway bridge specification [85]

The Japanese design specification for highway bridges was revised after the 1995 Kobe earthquake due to the extensive damage of bridges. It is reported that liquefaction-induced lateral spreading was the main cause of bridge failure. As a result, guidelines were introduced to take into account the forces due to liquefaction-induced ground movement. The design idealization for liquefaction-induced forces is shown in Figure 6.31. The code advises practicing engineers to check the design of piles against bending failure according to the pressure distribution shown in Figure 6.32.

This pressure distribution was formulated by back-analyzing some of the piled bridge foundations of the Hanshin expressway that were not seriously damaged [144]. This check against lateral spreading forces is additional to the requirements against inertia. The background for such a design philosophy is illustrated in [143]. The author notes that when the ground moves, the force associated with the ground movement applies to a part of foundation in contact with the moving ground. The author argues that the

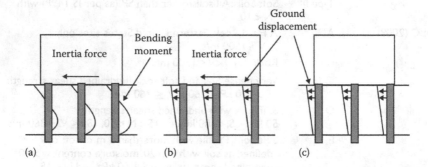

Figure 6.31 Schematic diagram showing the stages of pile failure during earthquake. (a) During shaking before liquefaction; (b) during shaking after liquefaction; (c) lateral movement after earthquake liquefaction. (From Choudhury et al., *Proceedings of the National Academy of Sciences*, India (Section A: Physical Sciences), Vol. 79, Pt. II, 1–11, 2009.)

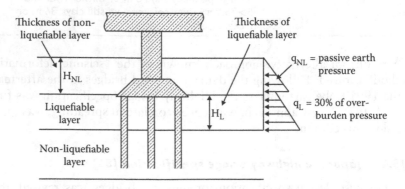

Figure 6.32 Idealization of seismic design of pile foundation in liquefied layer.

phenomenon is essentially a force mechanism and that it is appropriate to idealize the foundation as a structure supported by soil springs and prescribe the movement of ground at the end of each spring. In design of foundations based on such an analytical model, it is important to accurately predict the ground movement. Since the evaluation of maximum ground movement is difficult, the pressure distribution approach is incorporated in the code. Using dynamic centrifuge tests, [145] measured earth pressure acting on a piled foundation behind a retaining wall during and after earthquake loading. The authors concluded that the code in [85] over-predicts the lateral pressure in both liquefiable and non-liquefiable layers. The centrifuge test results of [146] show similar order of magnitude pressure, as predicted by [85] code. Dynamic centrifuge tests were performed in [147] and the results show that [85] under-predicts the lateral pressure distribution during the peak transient phase but gives a reasonable prediction for the post-earthquake residual flow.

6.13.2.3 Eurocode 8 (1998)

The Eurocode 8 (Part 5, 1998) advises the designers to design piles against bending due to both inertia and kinematic forces arising from the deformation of the surrounding soil due to earthquake. Piles shall be designed to remain elastic. Kinematic interaction need only be considered for soil deposits (ground types D, S1,S2) [for details of ground type (See NEHRP (2009)] with consecutive layers of sharply contrasting stiffness and design acceleration and supporting structure of importance category III and IV. When this is not feasible, the sections of the potential plastic hinging must be designed according to the rules of Part 1–3 of Eurocode 8. In addition, Part 5 of Eurocode 8 (1988) allows that the potential plastic hinge shall be assumed for: (1) A region of 2d from the pile cap, (2) a region of $\pm 2d$ from any interface between two layers with markedly different shear stiffness (ratio of shear moduli > 6) where d denotes the pile diameter.

6.13.2.4 NEHRP (2000)

The code notes that an unloaded pile placed in the soil would be forced to bend similar to a pile supporting a building. The primary requirement is stability and is best provided by piles that can support their loads while still conforming to the ground motions, hence the need for ductility.

REFERENCES

1. Berril J.B., Christensen S.A., Kennan R.P., Okada W., Pettinga J.R. (2001): Case studies of lateral spreading forces on a piled foundation. *Geotechnique*, Vol. 51(6), pp. 501–517.

2. Madabhushi S. P. G., Patel D., Haigh S. K. (2005): Geotechnical aspects of the Bhuj Earthquake, in EEFIT Report on the Bhuj Earthquake, Institution of Structural Engineers, London.
3. Yamashita, K., Hamada, J., Onimaru, S., Higashino, M. (2012): Seismic behavior of piled raft with ground improvement supporting a base-isolated building on soft ground in Tokyo, *Soils and Foundations*, Vol. 52, pp. 1000–1015.
4. Kramer, S.L. (1996): *Geotechnical Earthquake Engineering*. Prentice-Hall, Upper Saddle River, NJ.
5. Burland, J.B. (1995): Closing ceremony, *Proceedings of 1st IS-Hokkaido '94*, 2, pp. 703–705.
6. Yasuda, S., Nagase, H., Oda, S., Masuda, T., Morimoto, I. (1994): A study on appropriate number of cyclic shear tests for seismic response analyses, *1st IS-Hokkaido '94*, 1, pp. 197–202.
7. Roscoe, K. H. and Schofield, A. N. (1963): Mechanical behavior of an idealized wet clay, *Second European Conference on Soil Mechanics*, Weisbaden, Germany, Vol. 1, pp. 47–54.
8. Roscoe, K. H. Burland, J. B. (1968): On the generalized stress-strain behavior of wet clay, In: J. Heyman and F. A. Leckie, eds., *Engineering Plasticity*, University Press, Cambridge, pp. 535–609.
9. Duncan, J. M. Chang, C. Y. (1970): Non-linear analysis of stress and strain in soils. *Journal of Soil Mechanics and Foundation Engineering Division*, *ASCE*, Vol. 96, No. SM5, pp. 1629–1653.
10. Hardin, B. O. and Drnevich, V. P. (1972): Shear modulus and damping ratio in soils: Measurement and parameter effects, *Journal of Soil Mechanics and Foundation Division, ASCE*, Vol. 98, No. SM6, pp. 603–624.
11. Hardin, B. O. (1978): The nature of stress-strain behaviour of soils, *Proceedings: Earthquake Engineering and Soil Dynamics*, ASCE, Pasadena, CA, Vol. 1, pp. 3–89.
12. Jamillkowski, M. and Loroueil, S., Lopresi, D.C.F (1991): Theme lecture: Design parameters from theory to practice, *Proceedings, Geo-Coast '91*, Yokohama, Japan, pp. 1–41.
13. Seed, H.B., and Idriss, I.M. (1970): Soil moduli and damping factors for dynamic response analysis. Report No EERC 70–10, Earthquake Engineering Research Centre, University of California Berkeley.
14. Zen, K., Umehara, Y., and Hamada, K. (1978): Laboratory tests and in-situ seismic survey on vibratory shear modulus of clayey soils with different plasticities, *Proceedings, Fifth Japan Earthquake Engineering Symposium*, Tokyo, pp. 721–728.
15. Kokushu, T., Yoshida, Y. and Esashi, Y. (1982): Dynamic properties of soft clay for wide strain range, *Soils and Foundations*, Vol. 22, No. 4, pp. 1–18.
16. Dobry, R. and Vucetic, M. (1987): Dynamic properties and seismic response of soft clay deposits, *Proceedings of International Symposium on Geotechnical Engineering of Soft Soils*, Mexico City, Vol. 2, pp. 1487–1520.
17. Sun, J.I., Golesorkhi, R. and Seed, H.B. (1988): Dynamic moduli and damping ratios for cohesive soils. EERC Report No. UCB/EERC-88/15.
18. Ohta, Y. and Goto, N. (1976): Estimation of s-wave velocity in terms of characteristics indices of soil, *Butsuri-Tanko*, Vol. 29, No. 4, pp. 34–41.

19. Seed, H.B., Wong, R.T., Idriss, I. M. and Tokimatsu, K. (1986): Moduli and damping factor for dynamic analyses of cohesionless soils, *Journal of Geotechnical Engineering, ASCE*, Vol. 112, No. GT11, pp. 1016–1032.

20. Imai, T. and Tonouchi, K. (1982): Co-relation of N value with s-wave velocity and shear modulus *Proceedings, 2nd European Symposium on Penetration Testing*, Amsterdam, pp. 57–72.

21. Rix, G.J. and Stokoe, K.H. (1991): Correlation of initial tangent modulus and cone penetration resistance, *Calibration Chamber Testing, International Symposium on Calibration Chamber Testing*, A.B. Huang, ed., Elsevier, New York, pp. 351–362.

22. Baldi, G., Bellotti, R., Ghionna, V., Jamiolkowsky, M., Marchetti, S. and Pasquelini, E. (1986): Flat dilatometer tests in calibration chambers, *Proceedings, In situ '86*, Geotechnical Special Publication 6, ASCE, New York, pp. 431–446.

23. Mayne, P.W. and Rix, G.J. (1993): Gmax.-q_c relationship for clays *Geotechnical Testing Journal*, ASTM, Vol. 16(1): 54–60.

24. Bettotti, R., Ghionna, V., Jamiolkowsky, M., Lancellotta, R and Manfredini, G. (1986): Deformation characteristics of cohesionless soils in insitu tests. *Proceedings, In situ '86, Geotechnical Special Publication 6*, ASCE, New York, pp. 47–73

25. Hryciw, R.D. (1990): Small strain shear modulus of soil dilatometer, *Journal of Geotechnical Engineering*, ASCE, 116(11): 1700–1716.

26. Bryne, P.M., Salgado, F. and Howie, J.A. (1991): G_{max} from pressuremeter test: theory, chamber test and field measurement. *Proceedings, 2nd International Conference on Recent Advances in Geotechnical Earthquake Engineering and Soil Dynamics*, St. Louis, MO, Vol. 1, pp. 57–63.

27. Govinda Raju, L., Ramana, G.V., Hanumantha Rao, C., and Sitharam, T.G. (2004): Site-specific ground response analysis. *Special section: Geotechnics and Earthquake Hazards, Current Science*, 87(10): 1354–1362.

28. Schnabel, P. B., Lysmer, J. and Seed, H. B. (1972): SHAKE: A computer program for earthquake response analysis of horizontally layered sites. EERC Report 72–12. Earthquake Engineering Research Center, Berkeley, California.

29. Gutenberg, B. and Richter, C.F. (1956): Earthquake magnitude intensity energy and acceleration (Second paper), *Bulletin of Seismological Society of America*, 46: pp. 104–145.

30. Seed, H.B., Idriss, I.M., and Kiefer, F.W. (1969): Characteristics of rock motion during earthquakes, *Journal of Soil Mechanics and Foundation Division*, ASCE, 95 No. SM5, September.

31. Naik, N.P. and Choudhury, D. (2014): Comparative study of seismic ground responses using DEEPSOIL, SHAKE and D-MOD for soils of Goa, India, In *Geo-Congress 2014*: Geotechnical Special Publication No. GSP 234, ASCE, edited by A. J. Puppala, P. Bandini, T. C. Sheahan, M. Abu-Farsakh, X. Yu and L. R. Hoyos, Reston, VA, USA, pp. 1101–1110, in CD ROM.

32. Schnabel, P. B., Seed, H. B. (1972): Accelerations in rock for earthquakes in Western United States. Reports No. EERC 72-2, University of California, Berkeley, July.

33. Rajendran, K., Rajendran, C.P., Thakkar, M. and Tuttle, M.P. (2001): The 2001 Kutch (Bhuj) earthquake: Coseismic surface features and their significance. *Current Science*, 80(11): 1397–1405.

34. Ishihara, K. (1985): Stability of natural deposits during earthquakes. *Proceedings, 11th International Conference on Soil Mechanics and Foundation Engineering*, San Francisco. Vol. 1, p. 321–376.

35. Youd, T. L.; Perkins, D. M. (1978): Mapping liquefaction: Induced ground failure potential. *Journal of the Geotechnical Engineering Division*, pp. 443–446.

36. Tuttle, L. W. M.; Charpentier, R. R.; Brownfield, E. M. (1999): The Niger Delta Petroleum System: Niger Delta Province, Nigeria Cameroon and Equatorial Guinea, Africa U.S. Geological Survey Open-File Report 99-50-H, Denver, Colorado, 70 pp.

37. Youd, T. L.; Idriss, I. M.; Andrus, R. D.; Arango, I.; Castro, G.; Christian, J. T.; Dobry, R; Finn, W. D. L.; Harder, L. F. Jr.; Hynes, M. E.; Ishihara, K.; Koester, J. P.; Liao, S. S. C.; Marcuson, W. F: I. I. I.; Martin, G. R.; Mitchell, J. K.; Moriwaki, Y.; Power, M. S.; Robertson, P. K.; Seed, R. B.; Stokoe; K. H. I. I. (2001): Liquefaction resistance of Soils: Summary Report from the 1996 NCEER and 1998 NCEER/NSF Workshops on Evaluation of Liquefaction Resistance of Soils. ASCE J. *Geotech. Geoenvironmental Eng.* 127(10): 817–833.

38. Wang, W. (1979): Some findings in soil liquefaction. Water Conservancy and Hydroelectric Power Scientific Research Institute, Beijing, China.

39. Seed, H.B., Idriss, I.M. and Arango, I. (1983): Evaluation of liquefaction potential using field performance data. *Journal of Geotechnical Engineering*, ASCE, 109(3): 458–482.

40. Seed, R. B., et al. (2003): Recent advances in soil liquefaction engineering: A unified and consistent framework, EERC-2003-06, Earthquake Engineering Research Institute, Berkeley, CA.

41. Bray, J. D., and Sancio, R. B. (2006): Assessment of the liquefaction susceptibility of fine-grained soils. *Journal of Geotechnical and Geoenvironmental Engineering*, 132, 1165–1177.

42. Boulanger, R. W., and Idriss, R. W. (2006): Liquefaction susceptibility criteria for silts and clays. *Journal of Geotechnical and Geoenvironmental Engineering.*, 132(11), 1413–1424.

43. Seed, H. B. and Idriss, I. M. (1971): Simplified procedure for evaluating soil liquefaction potential. *Journal of Soil Mechanics and Foundation Division*, ASCE, 97 (9): 1249–1273.

44. Seed, H.B. (1979): Soil liquefaction and cyclic mobility evaluation for level ground during earthquakes. *Journal of Geotechnical Engineering. Div.*, ASCE, 105(2): 201–255.

45. Seed H. B., Idriss I. M., Arongo I. (1982): Evaluation of liquefaction potential using field performance data. *Journal of Geotechnical Engineering. Div.*, ASCE, 109(3): 458–482.

46. Seed H. B., Tokimatso K., Harder L. F. (1985): The influence of SPT procedures in soil liquefaction resistance evaluation. *Journal of Geotechnical Engineering. Div., ASCE*, 111(12).

47. Robertson, P. K., and Wride, C. E. (1998): Evaluating cyclic liquefaction potential using cone penetration test. *Canadian Geotechnical Journal*, 35 (3): 442–459.

48. Seed, R. B., Cetin, K. O., Moss, R.E.S, Kammerer, A. M., We, J., Pestana, J. M. and Riemer, M. F. (2001): Recent advances in soil liquefaction engineering and seismic site response evaluation Paper No. SPL-2. http://nisee.berkeley.edu/.

49. Cetin, K.O., Seed, R.B., Kiureghain, A.D., Tokimatsu, K., Harder, L.F., Kayen, R.E., Moss, R.E.S. (2004): Standard penetration test-based probabilistic and deterministic assessment of seismic soil liquefaction potential. *Journal of Geotechnical and Geoenvironmental Engineering*, ASCE, 130 (12): 1314–1340.
50. Idriss, I.M., and Boulanger, R.W. (2004): Semi-empirical Procedures for evaluating liquefaction potential during earthquakes. *Proceedings of 11th International Conference on Soil Dynamics and Earthquake Eng., and 3rd International Conference on Earthquake Eng.*, D. Doolin et al., eds., Stallion Press, Vol 1, 32–56.
51. Idriss, I.M., and Boulanger, R.W. (2004): Soil liquefaction during earthquakes, Earthquake Engineering Research Institute, Monograph MNO-12.
52. Idriss, I. M., and Boulanger, R. W. (2008): Soil liquefaction during earthquakes. Monograph MNO-12, Earthquake Engineering Research Institute, Oakland, CA, 261
53. Boulanger, R. W., Kamai, R., and Ziotopoulou, K. (2014): Liquefaction induced strength loss and deformation: Simulation and design. *Bulletin of Earthquake Engineering*, 12: 1107–1128.
54. Mhaske S. Y. and Choudhury D., (2010): GIS-based soil liquefaction susceptibility map of Mumbai city for earthquake events, *Journal of Applied Geophysics*, 70(3), 216–225.
55. Choudhury D., Phanikanth, V.S., Mhaske, S.Y., Phule, R.R. and Chatterjee, K. (2014): Seismic liquefaction hazard and site response for design of piles in Mumbai city. *Indian Geotech. J.*
56. Boulanger, R. W., and Idriss, I. M. (2004): Evaluating the potential for liquefaction or cyclic failure of silts and clays. Report No. UCD/CGM-04/01, Center for Geotechnical Modeling, Dept. of Civil and Environmental Engineering, Univ. of California, Davis, CA.
57. Brabharan, P., (2000): Earthquake ground damage hazard studies and their use in risk management in the Wellington Region, New Zealand. 12th World Conference on Earthquake Engineering, Auckland, New Zealand. Paper No. 1588.
58. Piya, B.K. (2004): Generation of a geologic database for the liquefaction hazard assessment in Kathmandu valley. M.Sc. Dissertation report. International Institute for Geoinformation Science and Earth Observation, Enschede, the Netherlands.
59. Baise, L. G., Higgins, R. B., and Brankman, C. M. (2005): Liquefaction hazard mapping: Statistical and spatial characterization of susceptible units. *Journal of Geotechnological and Geoenvironmental Engineering*, 1326, 705–715.
60. Pearce, J. T. and Baldwin, J. N. (2005): Liquefaction susceptibility mapping, St. Louis, Missouri and Illinois. Final Technical Report, U.S. Geological Survey, National Earthquake Hazards Reduction Program.
61. Bol, E., Onalp, A. and Ozocak, A. (2008): Liquefiability of silts and the vulnerability map of Adapazari. The 14th World Conference on Earthquake Engineering, Beijing, China.
62. Knudsen, K. L., Bott, J.D.J., Woods, M.O., and McGuire, T.L. (2009): Development of a liquefaction hazard screening tool for Caltrans bridge sites. *Proceedings of the 2009 ASCE Technology Council on Lifeline Earthquake TCLEE Conference*, Oakland, 573–584.

63. Li, X., Li, G., Li, C. (2009): Analysis of ground motion parameters and ground liquefaction prediction using GIS for Kunming Basin, China. *International Conference on Geo-Spatial Solution for Emergency Management and the 50th Anniversary of the Chinese Academy of Surveying and Mapping*, Beijing, China, 78–83.

64. Menezes, G. B., Galvao, T. C. B. (2009): Geostatistical assessment of liquefaction models: a GIS-based approach. 10th International Symposium on Environment Geotechnology and Sustainable Development, Bochum.

65. Mostafa, H., Baise, L., Hafez, H. and Zahaby, K. L. (2010): A GIS-based assessment of liquefaction potential of the city of New Damietta, Egypt. Geophysical Research Abstracts, 12, EGU 2010–15630.

66. Karakas, A., Coruk, O. (2010): Liquefaction analysis of soils in the Western Izmit Basin, Turkey. *Environment and Engineering Geoscience*, 16(4): 411–430.

67. Mote, T.I. and Dismuke, J.N. (2011): Screening-level liquefaction hazard maps for Australia. Australian Earthquake Engineering Society 2011 Conference, Barossa Valley, South Australia.

68. Tosun, H., Seyrek, E., Orhan, A., Savas, H. and Urk, M. T. (2011): Soil liquefaction potential in Eskisehir, NW Turkey. *Natural Hazards Earth System and Science*, 11, 1071–1082.

69. Habibullah, B.M., Pokhrel, R.M., Kuwano, J. and Tachibana, S. (2012): GIS-based soil liquefaction hazard zonation due to earthquake using geotechnical data. *International Journal of Geomatics*, 2(1): 154–160.

70. Hamada, M. (1992a): Large ground deformations and their effects on lifelines: 1964 Niigata earthquake, Technical Report NCEER 92-001, Chapter 3 of Hamada, M. and O'Rourke, T.D., National Center for Earthquake Engineering Research, 1–123.

71. Ishihara, K. (1997): Geotechnical aspects of the 1995 Kobe Earthquake, Terzaghi Oration, Proceedings of 14th International Conference on Soil Mechanics and Foundation Engineering, Hamburg, Germany, 2047–2073.

72. Tokimatsu, K. and Asaka Y. (1998): Effects of liquefaction induced ground displacements on pile performance in the 1995 Hyogoken-Nambu Earthquake, *Soils and Foundations Special Issue*: 163–177.

73. Finn, W.D.L., Thavaraj, T. and Fujita, N. (2001): Piles in liquefiable soils: Seismic analysis and design issues, *Proceedings of 10th International Conference on Soil Dynamics and Earthquake Engineering*, 4, 169–178.

74. Abghari, A. and Chai, J. (1995): Modelling of soil–pile–superstructure interaction for bridge foundations, *Proceedings of Performance of Deep Foundations under Seismic Loading*, J. P. Turner ed. ASCE, New York, 45–59.

75. Tabesh, A. and Poulos, H.G. (2001): Pseudo static approach for seismic analysis of singlepiles, *Journal of Geotechnical and Geoenvironmental Engineering*, ASCE, 127, 757–765.

76. Liyanapathirana D.S., Poulos H.G. (2005): Pseudo-static approach for seismic analysis of piles in liquefying soil. *Journal of Geotechnical and Geoenvironmental Engineering* ASCE 131, 1480–1487.

77. Phanikanth, V.S., Choudhury, D. and Reddy, G.R. (2013): Behavior of single pile in liquefied deposits during earthquakes, *International Journal of Geomechanics, ASCE*, 13(4), 454–462.

78. Hetenyi M. (1955): *Beams on Elastic Foundation: Theory with Applications in the Fields of Civil and Mechanical Engineering.* Ann Arbor: The University of Michigan Press.
79. Chatterjee, K., Choudhury, D., Poulos, H.G. (2015): Seismic analysis of laterally loaded pile under influence of vertical loading using finite element method, *Computers and Geotechnics*, Vol. 67, pp. 172–186.
80. JRA (2002): Specification for Highway Bridges, Part V. Japanese Road Association.
81. Liu, L. and Dobry R. (1995): Effect of liquefaction on lateral response of piles by centrifuge tests, NCEER report to Federal Highway Administration (FHWA). NCEER Bulletin, January, 9(1).
82. Tokimatsu, K., Suzuki, H. and Sato, M. (2004): Effects of inertial and kinematic forces on pile stresses in large shaking table tests. *Proceedings of the 13th World Conference on Earthquake Engineering*, Vancouver, Canada, Paper No. 1322.
83. Tokimatsu, K., Suzuki, H. and Sato, M. (2005): Effects of dynamic soil pile structure interaction on pile stresses, *Journal of Structural and Constructional Engineering*, 87: 125–132.
84. Tokimatsu, K. and Suzuki, H. (2005): Effect of inertial and kinematic interactions on seismic behavior of pile foundations based on large shaking table tests. *Proceedings of the 2nd CUEE Conference on Urban Earthquake Engineering*, Tokyo Institute of Japan, Japan.
85. JRA (1996): Specification for Highway Bridges, Part V. Japanese Road Association.
86. Puri, V.K. and Prakash, S. (2008): Pile design in liquefying soil. 14th World Conference on Earthquake Engineering (14WCEE-2008), October 12–17, 2008, Beijing, China.
87. Dobry, R., Abdoun, T, O'Rourke, T.D and Goh, S.H. (2003): Single piles in lateral spreads: Field bending moment evaluation. *Journal of Geotechnical and Geoenvironmental Engineering*, 129: 879–889.
88. JRA (1980): Japanese Road Association, Specification for Highway Bridges, Part V, Japanese Road Association.
89. Dobry, R., Taboda,V. and Liu L. (1995): Centrifuge modeling of liquefaction effects during earthquakes, *Proceedings of 1st International Conference on Earthquake Engineering*, Tokyo, Vol.3, pp.1291–1324.
90. Ashford, S. and Juirnarongrit, T., (2004): Evaluation of force based and displacement based analysis for response of single piles to lateral spreading, *Proceedings of 11th International Conference on Soil Dynamics and Earthquake Engineering and 3rd International Conference Earthquake Geotechnical Engineering*, University of California, Berkeley, Vol. 1, pp. 752–759.
91. API (2003): Recommended practice for planning designing and constructing fixed offshore platforms, American Petroleum Institute.
92. Architectural Institute of Japan (2001): Recommendation for design of building foundation. (In Japanese).
93. Tokimatsu, K., Oh-Oka, H., Satake, K., Shamoto, Y., and Asaka, Y. (1988): Effects of lateral ground movements on failure pattern of piles in the 1995 Hyogoken-Nambu Earthquake *Proceedings of Geotechnical Earthquake Engineering and Soil Dynamics III*, Reston, pp. 1175–1186.

94. Tokimatsu, K. (1999): Performance of pile foundations in laterally spreading soils. *Proceedings of 2nd International Conference on Earthquake Geotechnical Engineering*, 957–964, Lisbon, Portugal.

95. Ishihara, K. and Cubrinovski, M. (1998): Performance of large-diameter piles subjected to lateral spreading of liquefied depoits, Thirteenth Southeast Asian Geotechnical Conference, Taipei, Taiwan.

96. Hamada, M., Yasuda, S., Isoyama, R. and Emoto, K. (1986): Study on liquefaction induced permanent ground displacements, Association for the Development of Earthquake Prediction, Tokyo, Japan.

97. Shamato, Y., Zhang, J.M. and Tokimatsu, K. (1998): Methods for predicting residual postliquefaction ground settlement and lateral spreading. *Soils and Foundations* 38: 69–83.

98. Meera, R.S., Shanker, K. and Basudhar, P.K. (2007): Flexural response of piles under liquefied soil conditions, *Geotechnical and Geological Engineering*, 25, 409–422.

99. Hamada, M. (1992b). Large ground deformations and their effects on lifelines: 1964 Niigata earthquake, Technical Report NCEER 92-001, chapter 3 of Hamada, M. and O'Rourke, T.D., National Center for Earthquake. Engineering Research, 1–123.

100. Tokimatsu, K., Mizuno, H., Kakurai, M. (1996): Building damage associated with geotechnical problems, *Soils and Foundations*, pp. 219–234 Special issue on Geotechnical Aspects of the January 17 1995 Hyogoken-Nambu Earthquake.

101. Soga (1997): Geotechnical aspects of Kobe earthquake. EEFIT report on the Kobe Earthquake, Institution of Civil Engineers, UK.

102. Elahi, H.; Moradi, M.; Poulos, H.G.; Ghalandarzadeh, A. (2010): Pseudostatic approach for seismic analysis of pile group. *Computers and Geotechnics*, 37: 25–39.

103. Mindlin RD. (1936): Force at a point in the interior of a semi-infinite solid. *Physics*, 7:195–202.

104. Luco, J.E. (1982): Linear soil: Structure interaction: A review. *Applied Mechanical Division*, 53, 41–57.

105. Dobry, R.; Gazetas, G. (1988): Simple method for dynamic stiffness and damping of floating pile groups. *Geotechnique*, 38(4): 557–574.

106. Roesset, J.M.; Stokoe, K.H.; Baka, J.E.; Kwok, S.T. (1986): Dynamic response of vertical loaded small-scale piles in sand. *Proceedings on 8th European Conference on Earthquake Engineering*, Lisbon, 2: 5.6/65–72.

107. Banerjee, P.K. and Sen, R. (1987): Dynamic behavior of axially and laterally loaded piles and pile groups. Chapter 3 in *Dynamic Behavior of Foundations and Buried Structures*, Elsevier App. Sc., London, 95–133.

108. Wolf, J.P. (1988): *Soil: Structure Interaction Analysis in Time Domain*. Englewood Cliffs, NJ: Prentice-Hall.

109. Gazetas, G.; Makris, N. (1991): Dynamic pile–soil–pile interaction. I: Analysis of axial vibration. *Journal of Earthquake Engineering and Structural Dynamics* 20: 2.

110. Finn, W.D.L.; Wu, G; Thavaraj, T. (1997): Soil: Pile structure interaction. Geotechnical special publication, ASCE, No. 70: 1–22.

111. Naik, N.P. and Choudhury, D. (2014): Development of fault and seismicity maps for the state of Goa, India, *Disaster Advances*, Vol. 7, No. 6, pp. 12–24.

112. Shukla, J. and Choudhury, D. (2012): Estimation of seismic ground motions using deterministic approach for major cities of Gujarat, *Natural Hazards and Earth System Sciences*, Vol. 12, pp. 2019–2037.
113. Desai, S. and Choudhury, C. (2015): Site-specific seismic ground response study for nuclear power plants and ports in Mumbai, *Natural Hazards Review*, (13): 04015001-1-04015002-13.
114. Tajimi, H, (1969): Dynamic analysis of a structure embedded in an elastic stratum, Procs IV WCEE, Chile.
115. Matlock, H.; Foo, S.; Tsai, C.; Lam, I. (1978): SPASM 8: A dynamic beam-column program for seismic pile analysis with support motion. Fugro, Inc.
116. Tokimatsu, K., Oh-Oka H, Satake, K., and Asaka, Y (1996): Effect of lateral ground movements on failure patterns of piles in the 1995 Hyogoken-Nambu Earthquake, *Proceedings on Geotechnical Earthquake Engineering and Soil Dynamics,* pp. 1175–1186.
117. Kumar, A., Choudhury, D., Shukla, J., Shah, D. L. (2015): Seismic design of pile foundation by using PLAXIS3D. *Disaster Advances*, 8(6), 33–42.
118. Choudhury D., Phanikanth V.S. and Reddy, G.R. (2009): Recent advances in analysis and design of pile foundations in liquefiable soils during earthquake: A review, *Proceedings of the National Academy of Sciences, India (Section A: Physical Sciences)*, Vol. 79, Pt. II, pp. 1–11.
119. Rao, V.D., Chatterjee, K. and Choudhury, D (2013): Analysis of single pile in liquefied soil during earthquake using FLAC3D. *Proceedings of the International Conference on "State of the Art of Pile Foundation and Pile Case Histories"* PILE – 2013, June 2–4, Bandung, Indonesia, Vol. 1, pp. F5-1–F5-7
120. Giannakos, S., Gerolymos, N. and Gazetas, G. (2011): Single pile vs. pile group lateral response under asymmetric cyclic loading, *The 4th Japan Greece Workshop on Seismic Design of Foundations, Innovations in Seismic Design and Protection of Cultural Heritage*, Special Issue: The 2011 East Japan Great Earthquake, Kobe, Japan, Vol. 3, Paper No. 52, 577–587.
121. Maiorano, R.M.S., Sanctis, L, Aversa, S and Mandolini, A. (2009): Kinematic response analysis of piled foundations under seismic excitation, *Canadian Geotechnical Journal*, 46, 571–584.
122. Knappett, J.A. and Madabhushi, S.P.G. (2009b): Influence of axial load on lateral pile response in liquefiable soils. Part II: Numerical modeling, *Geotechnique*, 59(7), 583–592.
123. Karthigeyan, S.; Ramakrishna, V.V.G.S.T.; Rajagopal, K. (2006): Influence of vertical load on the lateral response of piles in sand. *Computers and Geotechnics*, 33, 121–131.
124. Maheshwari, B.K.; Truman, K.Z.; El Naggar, M.H.; Gould, P. L. (2004): Three: Dimensional finite element nonlinear dynamic analysis of pile groups for lateral transient and seismic excitations. *Canadian Geotechnical Journal*, 41: 118–133.
125. Eslami, MM; Aminikhah, A; Ahmadi, MM (2011): A comparative study on pile group and piled raft foundations (PRF) behavior under seismic loading. *Computer Methods on Civil Engineering* 2, 185–199.
126. Knappett, J.A.; Madabhushi, S.P.G. (2012): Effects of axial load and slope arrangement on pile group response in laterally spreading soils. *Journal of Geotechnical and Geoenvironmental Engineering*, 138: 799–809.

127. Ishizaki, S.; Nagao, T.; Tokimatsu, K. (2011): Dynamic centrifuge model test of pile—Supported building with semi-rigid pile head connections in liquefiable soil. *The 4th Japan Greece Workshop on Seismic Design of Foundations, Innovations in Seismic Design and Protection of Cultural Heritage*, Special Issue: The 2011 East Japan Great Earthquake, Kobe, Japan, Vol. 2, Paper No. 23, 537–246.

128. Choudhury, D.; Shen, R.F.; Leung, C.F. (2008): Centrifuge model study of pile group due to adjacent excavation. In *GEOCONGRESS 2008: Characterization, Monitoring, and Modeling of GeoSystems*, Geotechnical Special Publication No. 179, ASCE, 141–148.

129. Brandenberg, S.J.; Boulanger, R.W.; Kutter, B.L.; Chang, D. (2005): Behavior of pile foundations in laterally spreading ground during centrifuge tests. *Journal of Geotechnical and Geoenvironmental Engineering*, 131: 1378–1391.

130. Abdoun, T., Dobry, R., O'Rouke, T.D. and Goh, S.H. (2003): Pile foundation response to lateral spreads: Centrifuge modeling. *Journal of Geotechnical and Geoenvironmental Engineering*, 129, 869–878.

131. Kang, M. A., Banerjee, S., Lee, F. H. and Xie, H. P. (2012): Dynamic soil–pile–raft interaction in normally consolidated soft clay during earthquakes. *Journal of Earthquake and Tsunami*, 6(4), 1250031-1–1250031-12.

132. Horikoshi K, Matsumoto T, Hashizume Y, Watanabe T and Fukuyama H (2003): Performance of piled raft subjected to dynamic loading. *International Journal of Physical Modelling*, 2, 51–62.

133. Gerolymos, N.; Gazetas, G. (2006): Winkler model for lateral response of rigid caisson foundations in linear soil. *Soil Dynamics and Earthquake Engineering*, 26: 347–361.

134. Novak M.; Nogami T.; Aboul-Ella F. (1978): Dynamic soil reactions for plane strain case. *Journal of Engineering Mechanics Division*, 104: 953–959.

135. Mondal, G. Prashant, A. Jain, Sudhir, S.K. (2012): Significance of interface nonlinearity on the seismic response of a Well-Pier system in cohesionless soil. *Earthquake Spectra*, 28(3): 1117–1145.

136. Veletsos A.S.; Wei Y.T. (1971): Lateral and rocking vibration of footings. *Journal of Soil Mechanics Foundation, Div*, 97: 1227–1248.

137. Veletsos AS, Verbic B. (1974): Basic response functions for elastic foundations. *Journal of Engineering Mechanics Division*, ASCE 100:189–202.

138. IS 1893 (2002): Indian Standard Criteria for Earthquake Resistant Design of Structures, Part 1-General provisions and buildings. Fifth Revision, Bureau of Indian Standard, New Delhi.

139. NEHRP (2009): Recommended Seismic Provisions for Buildings and Other Structures. Building Seismic Safety Council, 2009, prepared for the Federal Emergency Management Agency, Washington, DC.

140. ASCE 7 (2010): Minimum Design Loads for Buildings and Other Structures. American Society of Civil Engineers, Reston, VA.

141. JRA (1972): Japanese Road Association, Specification for Highway Bridges, Part V.

142. Yasuda, S. and Berrill, J.B. (2000): Observations of the earthquake response of foundations in soil profiles containing saturated sands, Proceedings of Geotechnical Engineering 2000: An International Conference on Geotechnical and Geological Engineering, 19–24 November 2000, Melbourne, Australia.

143. Kawashima, K. (2000): Seismic design and retrofit of bridges, 12WCEE, Paper No. 1818, Auckland, New Zealand, 2000.

144. Yokoyama, K., Tamura, K. and Matsuo, O. (1997): Design methods of bridge foundations against soil liquefaction and liquefaction-induced ground flow, Second Italy-Japan Workshop on Seismic Design and Retrofit of Bridges, Rome, Italy, Feb 27 and 28, 1997.

145. Sato, M., Ogasawara, M. and Tozah, T. (2001): Reproduction of lateral ground displacements and lateral flow earth pressures acting on a pile foundations using centrifuge modeling, Proceedings 4th International Conference on Recent Advances in Geotechnical Earthquake Engineering and Soil Dynamics and Symposium in honor of Professor W.D. Liam Finn, San Diego, CA, March, 26–31.

146. Dobry, R and Abdoun, T. (2001): Recent studies on seismic centrifuge modeling of liquefaction and its effect on deep foundation, Proceedings of 4th International Conference on Recent Advances in Geotechnical Earthquake Engineering and Soil Dynamics and Symposium in honor of Professor W. D. Liam Finn, San Diego, CA, March 26–31.

147. Haigh, S.K. (2002): Effects of earthquake-induced liquefaction on pile foundations in sloping grounds, PhD thesis, University of Cambridge, UK.

Chapter 7

Special foundations

This chapter covers selected special foundations. These include

- Geothermally activated foundation systems
- Reuse of foundations
- Caisson foundations
- Shaft foundations
- Offshore foundations

The successful planning, design, and construction of these special foundations require specialized knowledge and wide experience. This applies to the planners, authorization personnel, and constructors, as well as independent peer reviewers.

7.1 GEOTHERMALLY ACTIVATED FOUNDATION SYSTEMS

The use of geothermal energy in buildings is an environmentally friendly, sustainable practice. In Germany, the operation of buildings uses up to 40%–50% of all energy consumption. Therefore, geothermal energy offers a possibility to improve overall energy efficiency [1]. Geothermal energy describes the utilization of the subsoil and the groundwater to extract and to store thermal energy.

Geothermally activated foundation systems can consist of shallow as well as deep foundation elements. Recently, even massive retaining structures are geothermally activated as well [2,3]. Moreover, with the help of geothermal energy, traffic areas, like train platforms or bridges, can be kept free of ice and snow during the winter months [4,5].

Depending on the boundary conditions, different systems can be used to produce geothermal energy. These include geothermal sensors, surface collectors or the direct utilization of the groundwater. Additionally, massive foundation elements, like reinforcedconcrete piles, can be used as solid absorbers.

247

7.1.1 Physical basics

Geothermally activated foundation systems extract energy from subsoil that is warmer than the outside temperature in winter months. The energy is transferred via a heat pump into the building for heating. In the summer months, when the subsoil is cooler than the outside temperature, it is used to cool a building [6]. The principle of seasonal geothermal storage is shown in Figure 7.1.

The heat transfers occur in the direction of the lower temperature level. The transfer methods can be substance-related (conduction, convection, and dispersion), or substance-free (thermal radiation).

Moreover, a transport of heat can occur because of evaporation or condensation processes, frost and dew processes, pressure changes, and radioactive decay, as well as biological and chemical processes. Generally, when using geothermal energy, the substance-related transport mechanisms dominate, and the others can be neglected. The heat transfer mechanisms in the subsoil depend on the grain size and the degree of saturation [7].

Assuming that the subsoil is an incompressible, isotropic, homogenous, porous medium, the consideration of the heat balance can be calculated using Equation 7.1. The change of the internal energy of the subsoil per time unit is described in consequence of the heat transfer mechanisms conduction, convection, and dispersion, as well as internal heat sources. The derivation of the Equation refers according to [1].

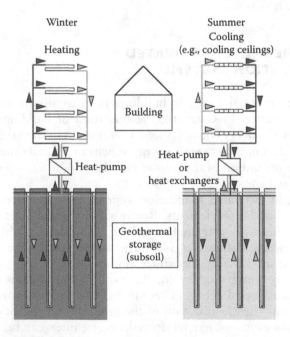

Figure 7.1 Seasonal geothermal storage.

$$\rho \cdot c \cdot \frac{\partial T}{\partial t} = div\left[\left(\lambda + (\rho \cdot c)_f \cdot \delta_\lambda \cdot |v|\right)gradT\right] - (\rho \cdot c)_f \cdot div(v \cdot T) + \dot{Q}_i \qquad (7.1)$$

where:
λ = thermal capacity [W/(mK)]
T = temperature [K]
t = time [s]
ρ = density [kg/m³]
c = specific thermal capacity [J/(kg·K)]
$\rho \cdot c$ = volumetric heat capacity [J/(m³·K)]
v = speed of fluid [m/s]
δ_λ = heat dispersivity [m]
$\dot{Q}_i$ = internal heat sources [W/m³]

7.1.2 Solid absorber

Solid absorbers are structurally required elements of foundation systems and retaining walls, which are geothermally activated. This includes, for example, energy piles, energy barrettes and geothermally activated foundation rafts. Recently, tunnel linings have been used for installing energy fleeces or energy segments [8–11].

For geothermal activation on the inside of the reinforcement cages of piles or in the blinding concrete of foundation rafts, heat exchanger tubes are installed in loops. Figure 7.2 shows a reinforcement cage with the attached heat-exchange pipes.

Figure 7.2 Heat exchanger tubes in a reinforcement cage. (From Katzenbach, et al., Geothermie, Ernst & Sohn Verlag, Berlin, 171–220, 2011.)

In principle, when using the geothermal energy, it should be ensured that the temperature of the construction elements is not below the frost level, because cyclic freezing and un-freezing reduces the bearing capacity of the soil.

7.1.3 Analysis and design

The determination of the number and the length, respectively, of the area of geothermally activated systems is determined on the basis of experience as well as analytical or numerical calculation [12]. In order to detect the long-term behavior as well as the extension of the influence of geothermal plants, numerical simulations (e.g., finite element method) are necessary. Thereby, the groundwater flow in particular has to be considered. In the case of high groundwater flow velocities, geothermal energy is transferred as a result of convective heat transfer processes [13].

Depending on the subsoil, an energy abstraction capacity of 40 W/m to 70 W/m can be indicated for energy piles. For piles with a diameter $D > 60$ cm, the energy abstraction capacity can be estimated to 35 W/m^2 of the surface. Geothermally activated foundation rafts have an energy abstraction capacity of approximately 15 W/m^2 [14].

Regarding stability and serviceability, geothermally activated foundation systems are assessed in the same way as non–geothermally activated systems. Moreover, the secured geothermal use, which means the sufficient energy abstraction or a sufficient energy storage in the subsoil, has to be checked. In any case it has to be guaranteed that there is no cyclic freezing and unfreezing to prevent a reduction of the bearing capacity of the subsoil.

Analogue to the Geotechnical Categories, geothermal constructions are classified into the Geothermal Categories GtC 1 to GtC 3, depending on the size and complexity, the ground conditions, and interactions with the surroundings [15].

The Geothermal Category GtC 1 includes small and simple geothermally activated constructions with an installed capacity of up to 30 kW in simple subsoil conditions. The construction is designed on experiences.

The Geothermal Category GtC 2 includes geothermal constructions with a medium size, with moderately difficult subsoil conditions that do not comply with the Geothermal Category GtC 1 or GtC 3.

The Geothermal Category GtC 3 includes complex geothermal constructions with an installed capacity of more than 100 kW and/or difficult subsoil conditions and complex interactions with the surroundings.

As a result of the requirements considering the design and construction of solid absorbers, these have to be classified into the Geothermal Category GtC 2.

With regard to the required quality assurance for planning, design, and construction of geothermally activated structures, more information is given in [1].

7.1.4 Construction

The construction of solid absorbers is the same as for non–geothermally activated foundations, taking into account the additional technical, physical and organizational conditions of the whole geothermal system.

Representative for the constructive design are the requirements of the structural design. The arrangement of heat exchanger tubes and measurement devices (e.g., temperature) is subsequently defined. The risk of damage to the additional installations must be minimized for a successful construction. For piles and barrettes, additional installations are fixed at the inside of the reinforcement cage as shown in Figure 7.2. For the reduction of the number of tubes, a series connection is useful. The head of an energy pile with two circuits, which are connected in series, is illustrated in Figure 7.3. Figure 7.4 provides a schematic view of the head of an energy pile. If reinforcedconcrete elements are directly in contact with the interior of a building, the heat exchanger tubes have to be located on the groundside of the structure.

Site connections, for example, at reinforcement cages of piles, should be avoided to keep the number of coupling joints as low as possible. In most cases, reinforcement cages have a maximum length of 15 m regarding the requirements of traffic and transport purposes. The difficulty of coupling segments of a reinforcement energy pile is shown in Figure 7.5.

Figure 7.3 Head of an energy pile. (From Katzenbach, et al., Geothermie, Ernst & Sohn Verlag, Berlin, 171–220, 2011.)

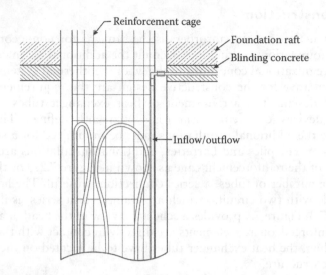

Figure 7.4 Scheme of a head of an energy pile.

Figure 7.5 Reinforcement joint of an energy pile. (From Katzenbach, et al., Geothermie, Ernst & Sohn Verlag, Berlin, 171–220, 2011.)

For reinforcedconcrete driven piles, the heat exchanger tubes are installed within the factory. The connections are protected by constructional measures during the ramming or vibrating process.

For geothermally activated foundation rafts, the heat exchanger tubes are placed on the blinding concrete. A fixing on a wide reinforcement grid is recommended.

Before operating the geothermal structures, the pipelines have to be flushed and the tightness has to be checked. This is followed by filling the pipelines with the heat transfer medium. Normally, water is the heat transfer medium for solid absorbers.

7.1.5 Examples from engineering practice

7.1.5.1 PalaisQuartier

This example from Frankfurt am Main, Germany, which is also described in Section 4.4, illustrates the utilization of foundation piles and piles of the retaining wall for seasonal geothermal energy storage [1–3,16]. The utilization of geothermal energy has been implemented for different projects in Frankfurt am Main, for example, at the high-rise structures Galileo, MainTower and Skyper [17]. For the construction project PalaisQuartier, 262 of the 302 foundation piles and 130 of the 289 piles of the retaining wall were installed as energy piles. A total of 392 energy piles are available for the development of geothermal resources of the subsoil with a total capacity of 913 kW. Figure 7.6 shows an energy pile, which is equipped with PE-lines and measuring cables. Figure 7.7 demonstrates the horizontal coupling at the base of the basement raft [18].

With the system nearly energy balanced, the subsoil is used as a seasonal thermal storage (Figure 7.8). In the heating period, the annual energy amount is about 2350 MWh/a. In the cooling period, the annual energy amount is about 2410 MWh/a.

For the water, legal and mining regulations, in addition to the verification of an intact energy balance, the geothermal impact were examined. The studies were carried out by three-dimensional numerical simulations considering the groundwater flow and the heat transfer (Figure 7.9).

As expected, the numerical simulations show the largest extensions of the influenced temperature field in the direction of the groundwater flow in the rocky Frankfurt Limestone. At a distance of about 25 m behind the retaining wall, the influence of the temperature of the subsoil decreases to $\Delta T < 1$ K. Particularly, the area with a concentration of piles shows the highest geothermal influences of the subsoil. The highest geothermal impact is at the office complex in the north of the whole building complex (Figure 7.10).

Figure 7.6 Heat exchanger tubes and measuring cables at the reinforcement cage. (From Katzenbach, et al., Geothermie, Ernst & Sohn Verlag, Berlin, 171–220, 2011.)

Figure 7.7 Horizontal connection of the heat exchanger tubes at the base of the basement raft. (From Katzenbach, et al., Geothermie, Ernst & Sohn Verlag, Berlin, 171–220, 2011.)

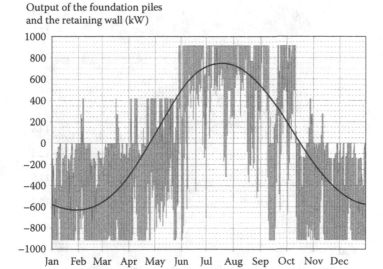

Figure 7.8 Energy-balanced seasonal storage. (From Katzenbach, et al., Geothermie, Ernst & Sohn Verlag, Berlin, 171–220, 2011.)

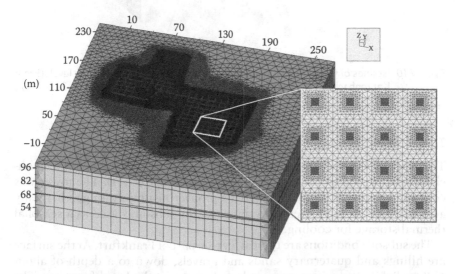

Figure 7.9 Three-dimensional FE model. (From Katzenbach, et al., Geothermie, Ernst & Sohn Verlag, Berlin, 171–220, 2011.)

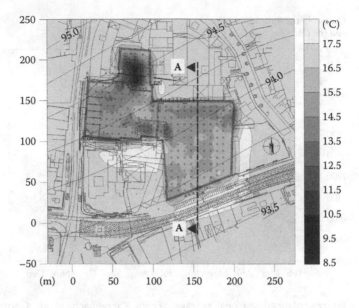

Cross-section A–A

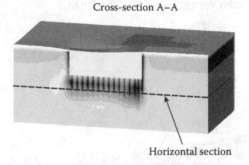

Horizontal section

Figure 7.10 Isolines of temperature after winter operation and groundwater level. (From Katzenbach, et al., Geothermie, Ernst & Sohn Verlag, Berlin, 171–220, 2011.)

7.1.5.2 Main Tower

The nearly 200 m tall high-rise building Main Tower is located in the financial district of Frankfurt am Main, Germany (Figure 7.11). The tower was constructed from 1996 to 1999 and is a pioneering project for geothermally activated foundation systems. In this case, the subsoil is used as a seasonal thermal storage for cooling.

The subsoil conditions are typical for this part of Frankfurt. At the surface are fillings and quaternary sands and gravels, down to a depth of about 6–8 m. Below the quaternary sands and gravels is the Frankfurt Clay. The Frankfurt Clay is a system of alternating layers of nearly non-permeable

Figure 7.11 Main Tower.

clay, permeable sand and permeable limestone. The limestone layers have a thickness of up to 2.8 m. The groundwater level is about 6 m below the surface [19].

The temperature of the subsoil and the groundwater is about 16°C at 30 m below the surface and increases with the depth.

For using the subsoil as seasonal heat storage, all 112 foundation piles as well as 101 reinforced piles of the retaining wall of the Main Tower were provided with heat exchanger tubes. Eight heat exchanger tubes were installed and distributed around the circumference of the foundation piles, which have a diameter of 1.5 m and a maximum length of 50 m below surface. The piles of the retaining wall, which have a diameter between 0.9 m and 1.5 m and a maximum length of 34 m below the surface, were equipped with four heat exchanger tubes. In total, 80,000 m heat exchanger tubes with a diameter of 25 mm were installed in the piles. Altogether, about 150,000 m³ of soil is available for storing the cooling energy [20].

In order to use the seasonal thermal storage of the Main Tower effectively, the subsoil has to be cooled down to the temperature level for summer operation. Therefore, a cold-air heat exchanger was installed on the roof of the high-rise building for winter operation (Figure 7.12). If the outside

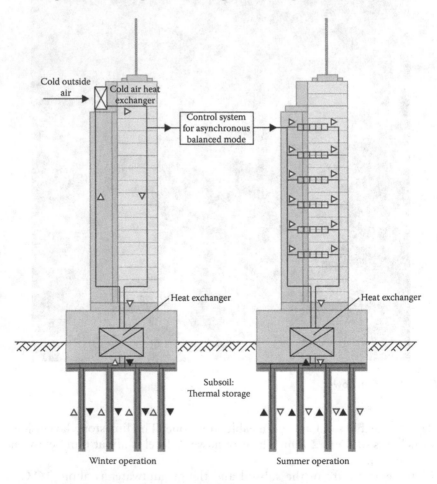

Figure 7.12 Seasonal thermal storage for cooling.

temperatures are below the temperatures inside, the geothermal storage is put into operation. The water circulating through the system is cooled by the cold-air heat exchanger in winter. In summer, the cool soil is used for cooling the building. In this case, the seasonal thermal storage is working in the asynchronous balanced mode.

7.2 REUSE OF FOUNDATIONS

For construction measures at existing structures, the reuse of existing foundations can be necessary for different kinds of reasons. These reasons include, for example:

- Increasing of the load due to conversion or additional storeys
- Constructions of excavations nearby or under existing foundations
- Irregular deformations, cracks
- Damage to the old foundations
- Changes in the subsoil
- Construction of a new structure using the existing foundation

7.2.1 Objectives of reuse

The deconstruction of existing foundations, particularly deep foundations, is time-consuming and causes considerable technical and economical difficulty. With deconstruction, a disturbed soil situation pertains, in which the new foundation components must be arranged [21].

Although the reuse of foundations has cost-saving potential, it can also generate new expense.

The reuse of existing foundation elements reduces construction time and completely eliminates the need for

- Planning and design of a new foundation system
- Construction of a new foundation system
- Deconstruction of existing foundations
- Disposal of excavation and demolition material

Conversely, the following extra expenses may occur:

- Investigation and evaluation of the existing foundation
- Strengthening of existing foundation elements
- If necessary, construction of additional foundation elements
- Connection to the superstructure

Due to continuously decreasing utilization cycles, space problems occur in inner cities, which determine the need for the construction of new foundation elements. Either there is generally not enough space to install the required equipment or machines, or the existing foundation elements are difficult obstacles [22–24].

Moreover, a reuse of existing foundations reduces the intervention in the subsoil. This is, for example, relevant for previously polluted sites or for sites of historic interest.

7.2.2 Geotechnical analysis

Reused foundations, maybe complemented with new foundation elements, have to be analyzed in the same way as new foundations elements regarding load-bearing capacity (ULS) and serviceability (SLS).

The alternatives for dealing with existing foundations are shown in Figure 7.13, exemplified for pile foundations. Either the existing foundation pile can be reused, maybe with a reduced load (Figure 7.13a), or additional piles have to be constructed, for example, in combination with a new foundation raft (Figure 7.13b). Alternatively, new piles and a traverse can be constructed (Figure 7.13c) to prevent the reloading of the existing pile. The last alternative is the partial or complete deconstruction of an existing pile and the construction of a new pile at the same place (Figure 7.13d).

To reuse existing foundations, comprehensive investigations are necessary. These also include investigations of the conditions and the functioning (integrity) of the internal and external bearing capacity. In this context, issues of serviceability should not be neglected [25].

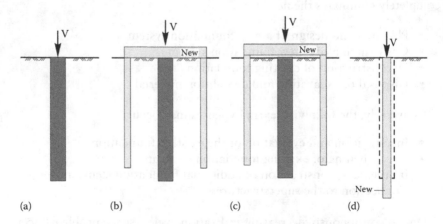

(a)   (b)   (c)   (d)

Figure 7.13 Alternatives of dealing with existing foundations.

7.2.3 Necessary investigations

The use of existing foundations bears a risk that is comparable to the subsoil risk. Through appropriate investigations in the planning phase, this risk can be reduced significantly. Further information would likely become apparent in the course of the construction works, which would then induce a new evaluation of the primary strategy [26].

Initially, all available information about the existing foundation elements has to be collected, examined, and evaluated. It has to be clarified, where the existing foundation elements are positioned and which dimensions they have. The required scope of the investigations depends on the quality of the received documents and information.

In order to clarify certain aspects, the following options are available:

- Nondestructive test methods
- Load tests
- Drill holes
- Digging out or pulling of a foundation element

The aim of the investigations is the determination of the permanent resistance for the verification of the load-bearing capacity (ULS) as well as a prognosis for the deformation behavior (SLS).

Indirect, nondestructive test methods for determining the integrity have to be calibrated [27]. For the investigation of foundation rafts, ultrasound echo acoustic methods and radar processes can be used. For the investigation of piles, the low strain method (hammer scale method) [28], the parallel seismic method, the mise-à-la-masse method, and the induction method can be used [29]. According to [30], load tests are advisable.

Examples of the reuse of existing foundation elements and the scope of necessary investigations are presented in [31–33].

7.2.4 Examples from engineering practice

7.2.4.1 Reichstag

The Reichstag building in Berlin, Germany, was constructed from 1884 to 1894. The building has a rectangular ground view with a width of 90 m and a length of 130 m. It has one sublevel. At each corner, a 36 m high tower with a ground area of 20 m×20 m is located. The building is founded on single and strip foundations consisting of limestone masonry. The transition between neighboring, high-strained, deep construction elements and higher, less-strained foundations was realized by a continuous changeover in foundation size in conjunction with reversed arches. The reversed arches consist of brick masonry, which were constructed on a limestone masonry capping (Figure 7.14). This construction method, in combination with the horizontal, compressive

Figure 7.14 Historical foundation of the Reichstag.

prestressing out of the arcs of the foundation masonry, results in a spatial distribution of the structural loads. The 160 MN heavy corner towers act as abutments to take the horizontal shearing force out of the reversed arches. Moreover, the corner towers make a substantial contribution to the whole building bracing. Due to soil conditions with a low stiffness, the tower in the northern corner and the dome are founded on about 3000 wooden piles. The piles have a diameter of 25 cm and a length of between 2.5 m and 5 m. The pile grid is about 1 m in square. The pile heads are embedded into a concrete grid.

The soil conditions are characterized by fine to medium-fine, silty sands (layer I). This layer is followed by coarse sands (layer II). The groundwater level is about 2.5 m below the surface [34].

After the reunification of Germany, the Reichstag was renovated, and the superstructure was altered in the early 1990s to host the German parliament [35]. Figure 7.15 shows the realized Reichstag. Notably, the old dome was replaced by a new glass dome, which was founded on 90 bored piles. The bored piles have a diameter of 0.9 m and 1.5 m and a length of between 15 m and 25 m. To reduce settlement, the piles were partly shaft-grouted and foot-grouted. In other parts, existing foundation elements such as the wooden piles have been reused. In the areas where the existing wooden piles could not be used, micro piles were installed or the area was improved by jet grouting.

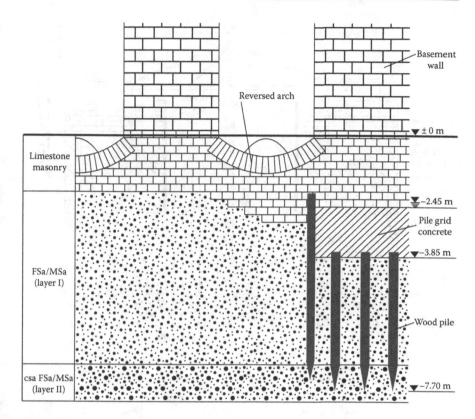

Figure 7.15 Reichstag.

Only fragmentary sketches and textual explanations in the building journals between 1884 and 1894 exist to give information about the existing foundation elements. Therefore, the foundational geometry had to be determined by a variety of test pits and core drillings.

In connection with the construction of the metro system, the groundwater was lowered about 10 m at the end of the 1930s. Due to the groundwater lowering, the environmental conditions for the wooden piles changed [36]. Special examination of the soil stratification and macroscopic and microscopic laboratory tests of the wooden piles were carried out to verify the reliability of these foundation elements. In addition, pile load tests on the wooden piles were carried out. The setup of the pile load test is shown in Figure 7.16. The results of the pile load tests are displayed in Figure 7.17. The bearing capacity of the wooden piles is between 200 kN and 300 kN in combination with a settlement of about 4 cm.

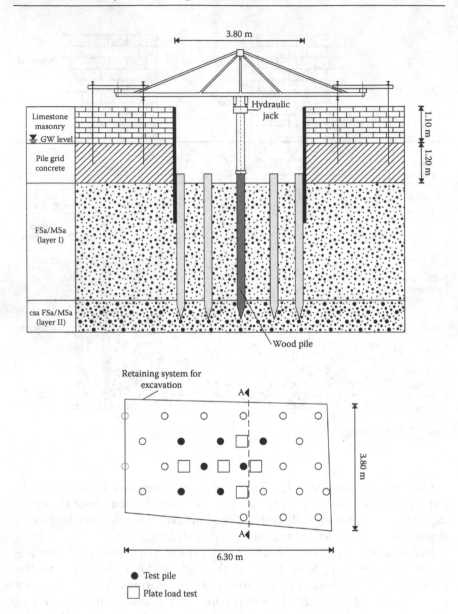

Figure 7.16 Setup of pile load tests and plate load tests.

In addition to the load tests on piles, load tests on reinforcedconcrete plates were made. The results of the plate load tests are displayed in Figure 7.18. The mobilized pressure under the plates varies between 630 kN/m^2 and 940 kN/m^2 in combination with a settlement of about 4 cm.

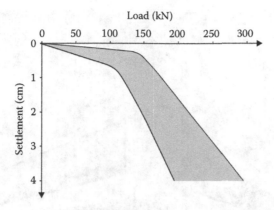

Figure 7.17 Results of pile load tests.

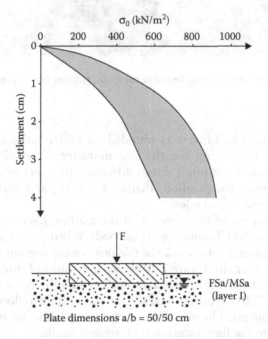

Plate dimensions a/b = 50/50 cm

Figure 7.18 Results of plate load tests.

7.2.4.2 Hessian parliament

The old plenary building of the Hessian parliament in Wiesbaden, Germany, was constructed in 1962. In 2008, it was replaced by a new building at the same place (Figure 7.19). For the purpose of resource conservation, it was planned to reuse the existing deep foundation elements [32].

Figure 7.19 New plenary building, founded partly on existing bored piles.

The old plenary building was founded on an irregular array of bored piles, which have varying lengths and diameters. The pile grid, which combines the individual piles, shows different cross-sections. The existing documents showed the position, diameter, length, and maximal bearing capacity of the individual piles.

The determination of the integrity of the existing piles was identified by the low strain method (hammer scale method). When detecting defects at the pile above the groundwater level, the pile heads were dug out and aborted in order to test them again. Figure 7.20 shows the results of three tests.

The expected length for the bored pile 1 with a diameter of 0.40 m was 4.30 m (Figure 7.20a). A destroyed section induced the reflection in a depth of 1 m below the piled head. On basis of the experimental results, the pile was cut down to the flaw location for testing it again.

The expected length for the bored pile 2 with a diameter of 0.40 m was 5.40 m (Figure 7.20b). The measured reflection from a depth of 5 m can be caused by the pile toe, which would indicate that the pile is too short, or by a destroyed section.

Figure 7.20c shows the test result of the bored pile 3. The expected length of 5 m was confirmed here.

According to the test results, about a quarter of the existing piles could be reused.

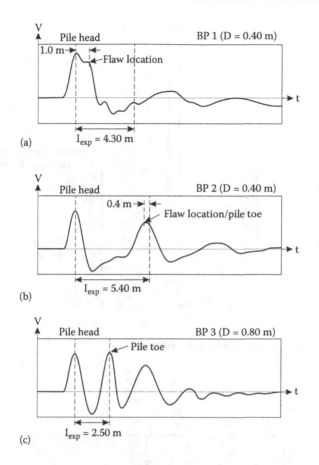

Figure 7.20 Results of integrity tests of existing bored piles.

7.3 SHAFT FOUNDATIONS

Shaft foundations are a simple form of open caisson foundations (cf. Chapter 7.5) and borrowed from well-digging. For installation, prefabricated reinforcedconcrete elements (e.g., shaft units) are inserted into the subsoil, while the soil is excavated in and below the elements. In consequence of the own weight and/or additional load, the elements sink into the subsoil. The geometry of the elements is normally circular, but it can also be elliptic or rectangular. The diameter can be chosen arbitrarily and generally varies between 1 m and 3 m. Figure 7.21 illustrates the construction. Where appropriate, a fluid that reduces the lateral friction can be used for a better sinking of the reinforcedconcrete elements.

When the final depth is reached, a reinforced (in simple cases not reinforced) raft is concreted. Then the hole is filled with concrete

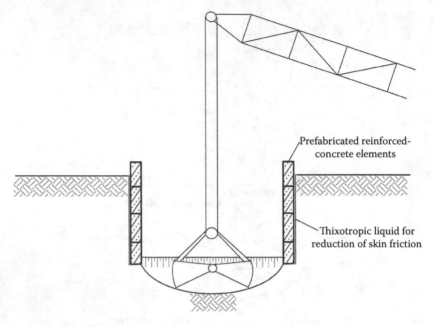

Figure 7.21 Shaft foundation under construction.

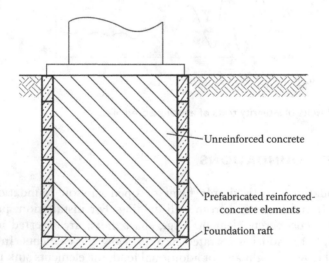

Figure 7.22 Completed shaft foundation.

(Figure 7.22). In special applications, reinforcement may be necessary depending on the project-related structural evidence (horizontal loads, low lateral bedding).

Shaft foundations may be considered as deep shallow foundations. Where appropriate, these analyses may be complemented by the analysis of safety against hydraulic failure or the safety against sinking. Consideration of shaft foundations as classic pile foundations has to be assessed in each individual case.

The advantage of shaft foundations is that no excavation is necessary to reach a soil layer with a sufficient strength and stiffness. The disadvantage, in comparison to, for example, deep foundations, is the construction tolerance. If shaft foundations are constructed below the groundwater level, a water load inside of the shaft foundation must be ensured to prevent a hydraulic failure. It is necessary to take a settlement trough in the vicinity into consideration, which occurs due to the construction process.

Based on their robustness, shaft foundations can also be used for special foundations. For example, they can be applied for bridge foundations on slopes [37].

7.4 CAISSON FOUNDATIONS

Caisson foundations were particularly constructed for foundations beneath the groundwater or seawater level [38]. They are generally reinforcedconcrete structures that are driven into the subsoil [39]. According to [40], caisson foundations are divided into open caisson foundations and air chamber caisson foundations. Caisson foundations can be designed as deep founded spread foundations. Where appropriate, the design has to consider the analysis of safety against hydraulic failure or sinking. Consideration of a caisson foundation as a classic pile foundation has to be assessed in each individual case. It is necessary to take into consideration a settlement trough in the area of a caisson foundation.

Open caisson foundations and air chamber caisson foundations can be used for bridge piers, lighthouses, and tunnel segments, as well as for water wings [40–42]. The geometry and dimensions of caissons can be freely selected.

7.4.1 Open caisson foundations

In open caisson foundations, the excavation base is accessible from the top (Figure 7.23). To guarantee the safety against hydraulic failure, a corresponding water load is required while constructing the caisson foundation under the water level. In comparison to shaft foundations, open caisson foundations are complex, reinforcedconcrete elements with a specially adapted geometry.

To create a surcharge in an open caisson foundation, large efforts are needed. Therefore, to avoid these, a comparatively high own weight is often necessary. A comparatively high own weight is necessary. In comparison to air chamber caisson foundations, open caisson foundations create bigger settlements and bigger deviations from the desired position.

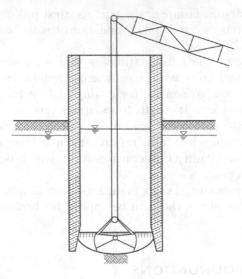

Figure 7.23 Open caisson foundation.

7.4.2 Air chamber caisson foundations

In air chamber caisson foundations, the excavation occurs in an isolated working chamber, which is linked with a sluice at the air side (Figure 7.24). The air pressure displaces the water inside the working chamber. Regarding a possible pressure failure, the sluice is installed at a height above a possible inner water line.

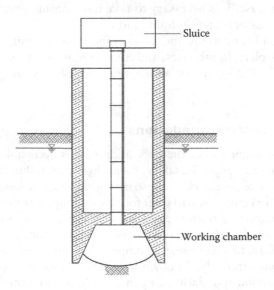

Figure 7.24 Air chamber caisson foundation.

7.5 OFFSHORE FOUNDATIONS

Offshore foundations are applied for temporary (drilling or working platforms) or durable (wind energy plants) constructions. In the context of the development of renewable energy, offshore wind energy plants are increasingly of interest. Offshore foundations are a special subject of marine geotechnical engineering and are still the object of extensive research and development activity [43–58]. Different offshore foundations have been developed, for example:

- Steel framework towers on pile foundations
- Tripod foundations
- Tripile foundations
- Monopile foundations
- Gravity foundations

An overview of the different systems is given in [41] and [59]. Figure 7.25 illustrates a wind energy plant that is founded on a tripile foundation in the North Sea at Hooksiel, close to Wilhelmshaven, Germany.

Figure 7.25 Offshore wind turbines with Tripile foundation.

Due to the complexity of planning, design, construction and monitoring of offshore foundations, a detailed treatment is not given here. Instead, references are given to the relevant expert literature in the following Chapter 7.6.

REFERENCES

1. Katzenbach, R.; Clauß, F.; Waberseck, T.; Wagner, I.M. (2011): Geothermie. Beton-Kalender 2011, Ernst & Sohn Verlag, Berlin, 171–220.
2. Fuller, R.M.; Hoy, H.E. (1970): Pile load tests including quick-load test method, conventional methods and interpretations. Research Record HRB 333, Highway Research Board, Washington, DC, 74–86.
3. Katzenbach, R.; Vogler, M.; Waberseck, T. (2008): Große Energiepfahlanlagen in urbanen Ballungsgebieten. Bauingenieur 83, Springer VDI-Verlag, Düsseldorf, Germany, Heft 7/8, 343–348.
4. Katzenbach, R.; Waberseck, T. (2005): Innovationen bei der Nutzung geothermischer Energie im Verkehrswegebau. Bauingenieur 80, Springer VDI-Verlag, Düsseldorf, Germany, Heft 9, 395–401.
5. Katzenbach, R.; Waberseck, T. (2007): Nutzung von Erdwärme zur Beheizung von Bahnsteigen. Der Eisenbahningenieur, Eurailpress, Hamburg, Germany, Heft 1, 28–32.
6. Ennigkeit, A. (2002): Energiepfahlanlagen mit Saisonalem Thermospeicher. Mitteilungen des Institutes und der Versuchsanstalt für Geotechnik der Technischen Universität Darmstadt, Germany, Heft 60.
7. Farouki, O.T. (1986): Thermal properties of soils. *Series on Rock and Soil Mechanics*, Vol. 11, Trans Tech, Publications, Clausthal-Zellerfeld, Germany.
8. Pralle, N.; Franzius, N.; Gottschalk, D. (2009): Stadt Bezirk: Mobilität und Energieversorgung—Neue Synergiepotenziale am Beispiel geothermisch nutzbarer urbaner Tunnel. Bauingenieur 84, Sonderheft, Springer VDI-Verlag, Düsseldorf, Germany, 98–103.
9. Hofmann, K.; Schmitt, D. (2010): Geothermie im Tunnelbau: Konzept für die Nutzung der Geothermie am Beispiel des B 10 Tunnels in Rosenstein. Geotechnik, Ernst & Sohn Verlag, Berlin, Heft 2, 135–139.
10. Mayer, P.M.; Franzius, N. (2010): Thermische Berechnungen im Tunnelbau. Geotechnik, Ernst & Sohn Verlag, Berlin, Heft 2, 145–151.
11. Brandl, H.; Adam, D.; Markiewicz, R.; Unterberger, W.; Hofinger, H. (2010): Massivabsorbertechnologie zur Erdwärmenutzung bei der Wiener U-Bahnlinie U2. Österreichische Ingenieur- und Architekten-Zeitschrift ÖIAZ, Austria, Heft 7–12, 193–199.
12. Appel, S.; Kirsch, F.; Mittag, J.; Mahabadi, O.K.; Richter, Thomas. (2007): The design of large scale installations of geothermal plants: Case studies and simulation models. *Darmstadt Geotechnics*, Germany, No. 15, 129–147.
13. Van Meurs, G.A.M. (1986): Seasonal storage in the soil. Thesis, Departement of Applied Physics, University of Technology Delft, the Netherlands.
14. Verband Beratender Ingenieure (VBI). (2009): VBI-Leitfaden Oberflächennahe Geothermie. Band 18 der VBI-Schriftenreihe, 2. Auflage, Berlin.

15. Brandl, H.; Adam, D.: Die Nutzung geothermischer Energie mittels erdberührter Bauteile. Geotechnique, Vol. XL, 124–149.
16. Skempton, A.W. (1951): The bearing capacity of clays. *Building Research Congress*, September, London, 180–189.
17. Von der Hude, N.; Sauerwein, M. (2007): Practical application of energy piles. *Darmstadt Geotechnics*, Germany, No. 15, 111–127.
18. Katzenbach, R.; Clauss, F.; Waberseck, T.; Vogler, M.; Adamietz, U. (2007): Present developments in the field of energy pile and borehole heat exchanger systems. *Darmstadt Geotechnics*, Germany, No. 15, 149–175.
19. Katzenbach, R.; Moormann, Chr. (1988): Messtechnische Überwachung von Baugrube und Gründung des Hochhauses "Main Tower" in Frankfurt am Main. Messen in der Geotechnik '98, 19–20 February, Mitteilung des Instituts für Grundbau und Bodenmechanik, Technische Universität Braunschweig, Germany, Heft 55, 87–121.
20. Ennigkeit, A. (2002): Energiepfahlanlagen mit saisonalem Thermospeicher. Mitteilungen des Institutes und der Versuchsanstalt für Geotechnik der Technischen Universität Darmstadt, Germany, Heft 60.
21. Chapman, T.; Butcher, A.P.; Fernie, R. (2003): A generalised strategy for reuse of old foundations. *13th European Conference on Soil Mechanics and Geotechnical Engineering*, 25–28 August, Prague, Czech Republic, Vol. 1, 613–618.
22. Allenou, C. (2003): One careful owner. *Ground Engineering*, March, 34–36.
23. Chapman, T.; Marsh, B.; Foster, A.: Foundations for the future. ICE Civil Engineering 144, Paper 12340, 36–41.
24. St. John, H. (2000): Follow these footprints. *Ground Engineering*, 2000, 24–25.
25. Butcher, A.P.; Powell, J.J.M.; Skinner, H.D. (2006): *Re-use of Foundations for Urban Sites: A Best Practice Handbook*. HIS BRE Press, Bracknell.
26. Chapman, T.; Marcetteau, A. (2004): Achieving economy and reliability in piled foundation design for a building project. *The Structural Engineer 82*, Vol. 11, 32–37.
27. Briaud, J.-L.; Ballouz, M.; Nasr, G. (2002): Defect and Length Predictions by Nondestructive Testing (NDT) Methods for Nine Bored Piles. *International Deep Foundation Congress*, 14–16. February, Orlando, 173–192.
28. Kirsch, F.; Klingmüller, O. (2003): Erfahrungen aus 25 Jahren Pfahl-Integritätsprüfung in Deutschland: Ein Bericht aus dem Unterausschuss "Dynamische Pfahlprüfungen"des Arbeitskreises 2.1 "Pfähle"der Deutschen Gesellschaft für Geotechnik e.V., Bautechnik 80, Ernst & Sohn Verlag, Berlin, Heft 9, 640–650.
29. Taffe, A.; Katzenbach, R.; Klingmüller, O.; Niederleithinger, E. (2005): Untersuchungen an Fundamentplatten im Hinblick einer Wiedernutzung. Beton- und Stahlbetonbau 100, Ernst & Sohn Verlag, Berlin, Heft 9, 757–770.
30. Powell, J.J.M.; Butcher, A.P.; Pellew, A. (2003): Capacity of driven piles with time—Implications for re-use. *13th European Conference on Soil Mechanics and Geotechnical Engineering*, 25–28 August, Prague, Czech Republic, Vol. 2, 335–340.
31. Katzenbach, R.; Weidle, A.; Ramm, H. (2003): Geotechnical basics in modelling of the soil–structure interaction due to the sustainable re-use of historical foundations and structures. *International Conference on Reconstruction of Historical Cities and Geotechnical Engineering*, 17–19 September, St. Petersburg, Russia, Vol. 1, 85–94.

32. Katzenbach, R.; Ramm, H. (2006): Re-use of foundations in course of the reconstruction of the Hessian parliament complex: A case study. *International Conference on Re-use of Foundations for Urban Sites: RuFUS 2006*, 19–20 October, BRE Press Watford, UK, 385–394.
33. Katzenbach, R.; Ramm, H. (2006): Reuse of historical foundations. *International Conference on Re-use of Foundations for Urban Sites: RuFUS 2006*, 19–20 October, BRE Press, Watford, UK, 395–403.
34. Katzenbach, R.; Quick, H. (1996): Grundwassermanagement bei temporären Baumaßnahmen. Bauingenieur 71, VDI-Verlag, Düsseldorf, Germany, Heft 7/8, 297–304.
35. Maetzel, U. (1996): Der Umbau des Reichstagsgebäudes im Rahmen der Gesamtplanung des Deutschen Bundestages in Berlin. VDI-Berichte 1246, Berlin baut im Grundwasser, VDI-Verlag, Düsseldorf, Germany, 1–19.
36. Mönnich, H.-D. (1996): Die Verkehrsbauwerke für den zentralen Bereich Berlins: Vorrang für unterirdische Lösungen. VDI-Berichte 1246, Berlin baut im Grundwasser, VDI-Verlag, Düsseldorf, Germany, 21–64.
37. Brandl, H. (2009): Stützbauwerke und konstruktive Hangsicherungen. Grundbau-Taschenbuch, Teil 3: Gründungen und geotechnische Bauwerke. 7. Auflage, Ernst & Sohn Verlag, Berlin, 747–901.
38. Pulsfort, M. (2012): *Grundbau, Baugruben und Gründungen. Handbuch für Bauingenieure: Technik, Organisation und Wirtschaftlichkeit*. Springer Verlag, Heidelberg, Germany, 1568–1639.
39. Schmidt, H.G.; Seitz, J. (1998): *Grundbau*. Ernst & Sohn Verlag, Berlin.
40. Lingenfelser, H. (2001): Senkkästen. Grundbau-Taschenbuch, Teil 3: Gründungen. 6. Auflage, Ernst & Sohn Verlag, Berlin, 233–275.
41. De Gijt, J.G.; Lesny, K. (2009): Gründungen im offenen Wasser. Grundbau-Taschenbuch, Teil 3: Gründungen und geotechnische Bauwerke. 7. Auflage, Ernst & Sohn Verlag, Berlin, 355–425.
42. Hafenbautechnische Gesellschaft e.V.; Deutsche Gesellschaft für Geotechnik e.V. (2012): Empfehlungen des Arbeitsausschusses "Ufereinfassungen"Häfen und Wasserstraßen EAU 2012. 11. Auflage, Ernst & Sohn Verlag, Berlin.
43. Grabe, J.; Mahutka, K.-P.; Dührkop, J. (2005): Monopilegründungen von Offshore-Windenergieanlagen. Bautechnik 82, Ernst & Sohn Verlag, Berlin, Heft 1, 1–10.
44. Lesny, K. (2008): Gründungen von Offshore-Windenergieanlagen: Entscheidungshilfen für Entwurf und Bemessung. Bautechnik 85, Ernst & Sohn Verlag, Berlin, Heft 8, 503–511.
45. Achmus, M.; Kuo, Y.-S.; Abdel-Rahman, K. (2008): Zur Bemessung von Monopiles für zyklische Lasten. Bauingenieur 83, Springer VDI-Verlag, Düsseldorf, Germany, Heft 7/8, 303–311.
46. Achmus, M.; tom Wörden, F.; Müller, M. (2009): Tragfähigkeit und Bemessung axial belasteter Offshorepfähle. Pfahl-Symposium, 19–20. February, Braunschweig, Germany, Heft 88, 213–230.
47. Hartwig, U.; Miehe, A. (2012): Besonderheiten bei Flachgründungen für Offshore-Windenergieanlagen. mining geo, Nr. 1, 141–149.
48. Hartwig, U.; Mayer, T. (2012): Entwurfsaspekte bei Gründungen für Offshore-Windenergieanlagen. Bautechnik 89, Ernst & Sohn Verlag, Berlin, Heft 3, 153–161.

49. Hildebrandt, A.; Schlurmann, T. (2012): Wellenbrechen an Offshore Tripod-Gründungen: Versuche und Simulationen im Vergleich zu Richtlinien. Bautechnik 89, Ernst & Sohn Verlag, Berlin, Heft 5, 301–308.

50. Stahlmann, A.; Schlurmann, T. (2012): Kolkbildung an komplexen Gründungsstrukturen für Offshore-Windenergieanlagen. Bautechnik 89, Ernst & Sohn Verlag, Berlin, Heft 5, 293–300.

51. Cuéllar, P.; Baeßler, M.; Georgi, S.; Rücker, W. (2012): Porenwasserdruckaufbau und Bodenentfestigung um Pfahlrgründungen von Offshore-Windenergieanlagen. Bautechnik 89, Ernst & Sohn Verlag, Berlin, Heft 9, 585–593.

52. Henke, S.; Qiu, G.; Pucker, T. (2012): Spudcans als Gründungsform für Offshore-Hubplattformen. Bautechnik 89, Ernst & Sohn Verlag, Berlin, Heft 12, 831–840.

53. Rücker, W.; Lüddecke, F.; Thöns, S. (2012): Tragverhalten von Offshore Gründungskonstruktionen. Bautechnik 89, Ernst & Sohn Verlag, Berlin, Heft 12, 821–830.

54. Rudolph, C.; Grabe, J. (2013): Untersuchungen zu zyklisch horizontal belasteten Pfählen bei veränderlicher Lastrichtung. Geotechnik 36, Ernst & Sohn Verlag, Berlin, Heft 2, 90–95.

55. Rücker, W.; Karabeliov, K.; Cuéllar, P.; Baeßler, M.; Georgi, S. (2013): Großversuche an Rammpfählen zur Ermittlung der Tragfähigkeit unter zyklischer Belastung und Standzeit. Geotechnik 36, Ernst & Sohn Verlag, Berlin, Heft 2, 77–89.

56. Arshi, H.S.; Stone, K.J.L.; Gunzel, F.K. (2015): Cost efficient design of monopile foundations for offshore wind turbines. 16th European Conference on Soil Mechanics and Geotechnical Engineering, 13–17 September, Edinburgh, Scotland, 1237–1242.

57. Pop, C.; Zania, V.; Trimoreau, B. (2015): Numerical modelling of offshore pile driving. 16th European Conference on Soil Mechanics and Geotechnical Engineering, 13–17 September, Edinburgh, Scotland, 1351–1356.

58. Bertossa, A.D. (2015): Evaluating geotechnical uncertainty for Offshore Wind Turbine foundation design at St. Brieuc Wind Farm. 16th European Conference on Soil Mechanics and Geotechnical Engineering, 13–17 September, Edinburgh, Scotland, 1249–1254.

59. Katzenbach, R.; Boled-Mekasha, G.; Wachter, S. (2006): Gründung turmartiger Bauwerke. Beton-Kalender 2006, Ernst & Sohn Verlag, Berlin, 409–468.

Appendix A: ISSMGE Combined Pile-Raft Foundations Guideline (2013)

A.I TERMS AND DEFINITIONS

The Combined Pile-Raft Foundation (CPRF) is a geotechnical composite construction that combines the bearing effect of both foundation elements, raft and piles, by taking into account interactions between the foundation elements and the subsoil shown in Figure A.1.

The characteristic value of the total resistance $R_{tot,k}(s)$ of the CPRF depends on the settlement s of the foundation and consists of the sum of the characteristic pile resistances $\sum_{j=1}^{m} R_{pile,k,j}(s)$ and the characteristic base resistance $R_{raft,k}(s)$. The characteristic base resistance results from the integration of the settlement dependent contact pressure $\sigma(s,x,y)$ in the ground plan area of the raft (cf. Equations A.1 through A.3).

$$R_{raft,k}(s) = \iint \sigma(s,x,y)\,dxdy \tag{A.1}$$

$$R_{tot,k}(s) = \sum_{j=1}^{m} R_{pile,k,j}(s) + R_{raft,k}(s) \tag{A.2}$$

$$R_{pile,k,j}(s) = R_{b,k,j}(s) + R_{s,k,j}(s) \tag{A.3}$$

The bearing behavior of the CPRF is described by the pile–raft coefficient α_{CPRF}, which is defined by the ratio between the sum of the characteristic pile resistances $\sum_{j=1}^{m} R_{pile,k,j}(s)$ and the characteristic value of the total resistance $R_{tot,k}(s)$ (cf. Equation A.4):

$$\alpha_{CPRF} = \frac{\sum_{j=1}^{m} R_{pile,k,j}(s)}{R_{tot,k}(s)} \tag{A.4}$$

277

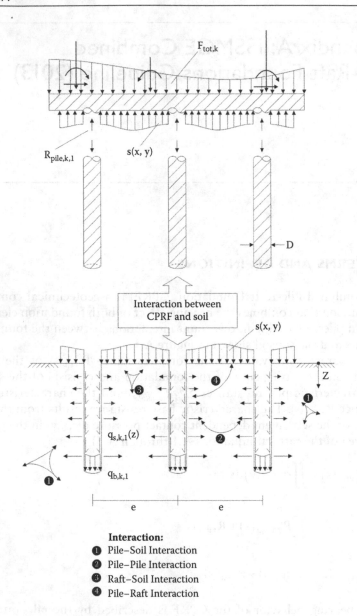

$F_{tot,k}$

$R_{pile,k,1}$

$s(x, y)$

D

Interaction between CPRF and soil

$s(x, y)$

z

$q_{s,k,1}(z)$

③

④

①

②

$q_{b,k,1}$

①

e   e

Interaction:
① Pile–Soil Interaction
② Pile–Pile Interaction
③ Raft–Soil Interaction
④ Pile–Raft Interaction

Figure A.1 Combined Pile-Raft Foundation (CPRF) as a geotechnical composite construction and the interactions coining the bearing behavior.

The pile–raft coefficient varies between $\alpha_{CPRF} = 0$ (spread foundation) and $\alpha_{CPRF} = 1$ (pure pile foundation). Figure A.2 shows a qualitative example of the dependence between the pile-raft coefficient α_{CPRF} and the settlement of a CPRF s_{CPRF} related to the settlement of a spread foundation s_{sf} with equal ground plan and equal loading.

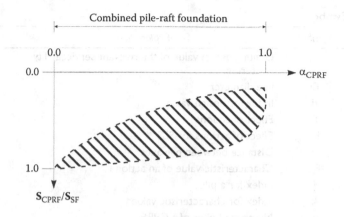

Figure A.2 Qualitative example of a possible settlement reduction of a CPRF in function of the pile-raft coefficient α_{CPRF}.

The pile-raft coefficient α_{CPRF} depends on the stress level and on the settlement of the CPRF.

A.2 SCOPE

The CPRF guideline applies to the design, dimensioning, inspection, and construction of preponderantly vertically loaded Combined Pile-Raft Foundations.

Note: The CPRF guideline can also be applied to deep foundation elements other than piles such as diaphragm walling elements (barrettes), diaphragm walls, sheet pile walls, and so on.

The CPRF guideline shall not be used in cases where layers of relatively small stiffness (e.g., soft, cohesive, and organic soils) are situated closely beneath the raft.

A.3 GEOTECHNICAL CATEGORY

According to Eurocode EC 7, the Geotechnical Category 3 may be assigned for the design of Combined Pile-Raft Foundation.

A.4 SYMBOLS

Symbols are given in Table A.1.

Table A.1 Symbols

Number	Symbol	Explanation	Unit
1	C_d	Limiting design value of the relevant serviceability criterion	
2	D	Pile diameter	m
3	d	Index for design value	—
4	E	Effect of actions	
5	E_2	Effect of actions for SLS	
6	e	Distance between pile axes	m
7	$F_{k,i}$	Characteristic value of an action i	MN
8	j	Index for a pile	—
9	k	Index for characteristic value	—
10	m	Number of piles of a CPRF	—
11	q_b	Unit base resistance	MN/m²
12	q_s	Unit shaft resistance	MN/m²
13	R	Resistance	MN
14	$R_{b,k}(s)$	Characteristic value of the base resistance of a pile as a function of settlement	MN
15	$R_{pile,k,j}(s)$	Characteristic value of the resistance of the pile j of a pile group as a function of settlement	MN
16	$R_{raft,k}(s)$	Characteristic value of the base resistance of a CPRF as a function of settlement	MN
17	$R_{s,k}(s)$	Characteristic value of the shaft resistance of a pile as a function of settlement	MN
18	$R_{tot,k}(s)$	Characteristic value of the total resistance of a CPRF as a function of settlement	MN
19	$R_{1,tot}$	Total resistance of a CPRF for ULS	MN
20	s	Settlement	m
21	s_{CPRF}	Settlement of a CPRF	m
22	s_{sf}	Settlement of a spread foundation	m
23	s_2	Allowable settlement for SLS	m
24	Δs_2	Allowable differential settlement for SLS	m
25	x,y,z	Cartesian coordinates	m
26	α_{CPRF}	Pile-raft coefficient	—
27	γ	Partial safety factor	—
28	γ_G	Partial safety factor for a permanent action	—
29	γ_Q	Partial safety factor for a variable action	—
30	γ_R	Partial safety factor for a resistance	—
31	$\sigma(s,x,y)$	Contact pressure as a function of settlement	MN/m²

A.5 SOIL INVESTIGATION AND EVALUATION

Soil investigation on site and in laboratory is required for the design and the dimensioning of a CPRF and is the basis for all analysis. The quality and quantity of the geotechnical investigations and the performance of the field and laboratory tests have to be designed and controlled by geotechnical experts and also have to be evaluated under the consideration of the soil–structure interaction.

The results of field and laboratory investigation have to be compared with values experienced for the local soil conditions.

A.5.1 Field investigation

Direct soil investigations are necessarily required for the design of a CPRF even if local experiences are given. Depending on project-related circumstances and the local soil conditions, the investigation program has to be reviewed concerning the necessity of further investigations.

A.5.2 Laboratory investigation

The design of a CPRF requires a sufficient knowledge of the deformation and the strength of the subsoil. Additional to classification tests, a sufficient number of laboratory tests on soil samples must be performed in order to determine the stiffness and shear strength of the soil. Quality and quantity of the laboratory tests have to be defined with regard to the constitutive laws used within the analysis of the CPRF.

A.5.3 Tasks within the construction process

Exposures during the construction process of a CPRF have to be examined and evaluated by a geotechnical expert, and they have to be compared to the results of the actual soil investigation. The data gathered during the construction of the bored piles have to be recorded in a protocol and displayed graphically by diagrams. The usage of driven piles or other deep foundation elements requires a corresponding procedure.

If the soil and groundwater conditions encountered during the construction process deviate relevantly from the expected soil and groundwater conditions, additional investigations of subsoil and groundwater have to be carried out. The updated geotechnical data is the basis for a reviewed design and construction process of the CPRF.

A.6 REQUIREMENTS TO THE COMPUTATIONAL METHODS FOR THE DESIGN OF A CPRF

A.6.1 Prefaces

The bearing effect of a CPRF is influenced by the interactions of the particular bearing elements (Figure A.1).

Beside the pile group effect, that is, the mutual interactions of the piles within the pile group, the contact pressure considerably influences the bearing behavior of the foundation piles of the CPRF.

Therefore, the prerequisite for a safe design of a CPRF is the realistic modeling of the interactions between the superstructure, the foundation elements, and the subsoil. This requires the use of a computational model that is able to simulate the interactions determining the bearing behavior of the CPRF in a reliable and realistic way.

The computational model used for the design of a CPRF shall contain a realistic geometric modeling of the foundation elements and the soil continuum as well as a realistic description of the material behavior of both structure and subsoil, and of the contact behavior between the soil and the foundation elements. The choice of the constitutive laws and the applied material parameters used within the analysis has to be justified.

A.6.2 Bearing behavior of a single pile

For the design of a CPRF, the knowledge of the bearing behavior of a standalone single pile under comparable soil conditions is required (Section A.6.3, Paragraph 1).

If no previous experience of the bearing behavior of a single pile by test loadings exists, a static pile test under axial loading has to be performed for a corresponding pile type under comparable soil conditions.

If no static load pile tests are performed, the bearing behavior of a single pile can be defined by using the empirical values indicated in the concerned standards. The transferability of the standardized empirical values on the soil conditions explored on-site and on the planned CPRF has to be proven.

A.6.3 Requirements for a computational model

The computational model shall be able to simulate the bearing behavior of an appropriate single pile according to Section A.6.2. The shearing at the pile shaft and the compression process at the pile base has to be modeled correctly.

The computational model used for the design of the CPRF shall also be able to transfer the bearing behavior of a single pile to the bearing behavior of the CPRF including the pile–pile interaction and the pile–raft interaction. Furthermore, the computational model has to be able to simulate all relevant interactions, including their effects on the bearing behavior of the CPRF (Figure A.1).

For the design of a CPRF, different computation methods are available, which are based on different computation and modeling approaches. The computation method used for the design of a CPRF has to be documented within the design process.

A.7 ULS: ULTIMATE LIMIT STATE

The proof of the external- and internal bearing capacity has to be carried out for a CPRF. The external bearing capacity describes the bearing capacity of the soil interacting with the foundation elements. The internal bearing capacity describes the bearing capacity of the single components like the piles and the foundations raft.

The bearing behavior of the CPRF is computed based on characteristic soil and material parameters. Time-dependent properties of the soil and the structure should be considered if necessary.

The stiffness of the superstructure and its influence on the bearing behavior of the CPRF has to be considered within the computational investigation and the proofs of limit states.

Figure A.3 shows the concept for the proof of ultimate limit state schematically.

A.7.1 Proof of the external bearing capacity (ULS)

A sufficient safety against failure of the overall system is achieved by fulfilling the following Equation A.5:

$$E_d = E_{G,k} \cdot \gamma_G + E_{Q,k} \cdot \gamma_Q \leq \frac{R_{1,tot,k}}{\gamma_R} = R_{1,tot,d} \tag{A.5}$$

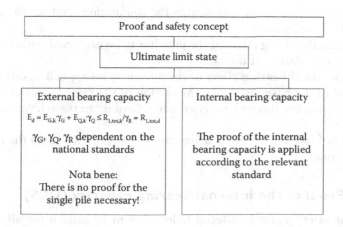

Figure A.3 Proof and safety concept in the ultimate limit state.

The characteristic value of the total resistance of the CPRF in the ultimate limit state $R_{1,tot,k}$ has to be determined by an analysis of the CPRF as an overall system based on a computational model including all relevant interactions according to Section A.6.2. The characteristic values of the soil and the structure properties shall be used within the analysis. The characteristic value of the total resistance $R_{1,tot,k}$ has to be derived from the load-settlement relation for the overall system. The characteristic value of the total resistance $R_{1,tot,k}$ is equal to the load at which the settlements of the CPRF visibly increase. In the load-settlement curve, the characteristic value of the total resistance $R_{1,tot,k}$ represents that point at which the flat section, after a transition region with increasing settlement, passes into the steeply falling section.

If the proof is not performed by a realistic computational model according to A.6.3 in simple cases, it is permissible to calculate the characteristic value of the total resistance $R_{1,tot,k}$ alternatively by means of the characteristic value of the base resistance of the foundation raft of the CPRF.

"Simple cases" are given if the following conditions are fulfilled:

- A geometrically uniform configuration of the CPRF:
  - Identical pile length and pile diameter.
  - Constant distance between the pile axes e.
  - Rectangular or round raft foundation.
  - Projection of the raft foundation beyond the outer pile row $\leq 3D$ (D = pile diameter).
- Homogeneous subsoil (no layering):
  - No distinct difference in stiffness between the individual layers.
- Actions
  - Centrically loaded raft foundation, that is, the resulting action is concentrated in the center of gravity of the raft.
  - No predominantly dynamic effects.

The bottom line of the raft defines the foundation level for the calculation of the base resistance.

The vertical bearing effect of the piles has to be neglected within the base resistance calculation of the raft.

The horizontal bearing effect of the piles may be applied as dowel resistance within the base resistance calculation of the raft. The calculation of the base resistance has to be carried out according to the relevant national standards.

The proof of the external bearing capacity of a CPRF saves the proof of all single piles.

A.7.2 Proof of the internal bearing capacity (ULS)

Sufficient safety against material failure has to be proven for all foundation elements according to the specific standards. The proof of the internal bearing capacity shall be carried out for all relevant combinations of actions. The following stress states have to be proven:

- Piles: Tension (construction stages), compression combined with bending and shearing.
- Raft: Bending, shearing, punching at the areas of punctual loading of the superstructure elements (columns) as well as of the foundation piles.

The calculation of the internal forces shall be performed for two cases because of the nonlinear relation between the settlement and the partial resistances of raft and piles. The pile-raft coefficient α_{CPRF} shall be calculated for both limit states, the ultimate limit state (Section A.7.1) and the serviceability limit state (Section A.8.1). The internal forces of the raft and the piles have to be computed due to the distribution of the characteristic actions on raft and piles determined by the pile–raft coefficient. The more unfavorable results have to be used for the design of the foundation elements.

The proof of the internal bearing capacity of the foundation elements has to be carried out according to the relevant standards.

If no detailed proof is performed, the piles have to be reinforced to the minimum amount or the amount calculated within the design process on their total length.

A.8 SLS: SERVICEABILITY LIMIT STATE

The proof of the serviceability limit state comprises two different examinations analogously to the proof of the ultimate limit state (Figure A.4).

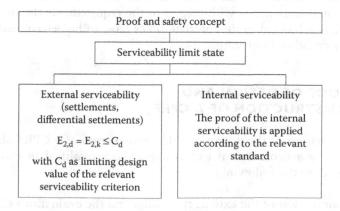

Figure A.4 Proof and safety concept in the serviceability.

A.8.1 Proof of the external serviceability

A sufficient safety of the serviceability is achieved by fulfilling the following Equation A.6:

$$E_{2,d} = E_{2,k} \leq C_d \tag{A.6}$$

The effects E dependent on the actions $F_{k,i}$ have to be computed by a computational model according to Section A.6.2 based on characteristic values for the material properties. The effects E are computed on the overall system subjected to nonfactorised actions.

During the service of the building, the effects E expressed by the relevant settlement s_2, differential settlements Δs_2, and so on have to be smaller than the limiting design value of the relevant serviceability criterion.

The value of the limiting design value of the relevant serviceability criterion C_d is defined by the requirements deriving from the characteristics of the planned CPRF and the adjacent buildings possibly affected by the construction of the CPRF. For the allowable settlements s_2 or the allowable differential settlements Δs_2, the limiting values need to be defined by taking into account the sensitivity of the structure for deformations and especially for differential settlements. It should also be checked for the sensitivity of the adjacent underground or overground structures and infrastructural installations.

A.8.2 Proof of the internal serviceability

For the foundation elements a sufficient safety for the serviceability limit state has to be proven according to the material specific standards. The following stress states have to be proven:

- Piles: Restriction of the crack width.
- Raft: Restriction of the crack width, allowable deflections and/or differential settlements with respect to the requirements the superstructure is subjected to the internal forces have to be determined for the serviceability limit state.

A.9 PROOF OF DESIGN AND CONSTRUCTION OF A CPRF

The examination of the design and the construction of a CPRF should be controlled by a geotechnical expert particularly qualified in this subject with respect to the following:

- Examination of the extent, the results and the evaluations of the soil investigation (field and laboratory tests).

- Evaluation of the plausibility and suitability of the characteristic values of the soil properties used in the computational models for the CPRF.
- Examination of the computational model used for the design of the CPRF and the computation results by using independent comparative calculations.
- Examination of the evaluation of the effects and the adjacent buildings.
- Examination of the measuring program and of the soil exposures attained within the construction process of the CPRF.
- Examination of the protocol of the acceptance procedure and the measured values.

A.10 CONSTRUCTION OF A CPRF

The construction of a CPRF has to be supervised by a geotechnical expert particularly qualified in this subject assigned by the owner or the supervising authority with respect to the ground engineering aspects. This applies to the construction both of the piles and the foundation level. The protocols of the acceptance procedure and the measured values have to be included in the examination.

A.11 MONITORING OF A CPRF

The bearing behavior and the force transfer within a CPRF may be monitored by a geotechnical expert particularly qualified in this subject due to the requirements deriving from the soil, the superstructure, and the foundation according to the concept of the observational method on the basis of the measuring program set up in the design phase. The monitoring comprises geotechnical and geodetic measurements at the new building and also at adjacent buildings. The monitoring of a CPRF is an elementary and indispensable component of the safety concept and is used for the following purposes:

- The verification of the computational model and the computational approaches.
- The in-time detection of possible critical states.
- An examination of the calculated settlements during the whole construction process.
- Quality assurance and the conservation of evidence both during the construction process and during the service of the building.

The monitoring program has to be designed by a geotechnical expert in the design phase. The measurements shall give information about the load distribution between the raft and the piles.

In simple cases, the arrangement and regular leveling of settlement measuring points can be sufficient.

BIBILIOGRAPHY

Bourgeois, E., Buhan, P. and Hassen, G. (2012), Settlement analysis of piled-raft foundations by means of multiphase model accounting for soil-pile interactions. *Computers and Geotechnics*, 46, 26–38.

Butterfield, R. and Banerjee, P.K. (1971), The problem of pile group–pile cap interaction. *Géotechnique*, 21(2), 135–142.

CEN European Committee of Standardization (2008), Eurocode 7: Geotechnical design—Part 1: General Rules.

Chaudhary, K. N., Phoon, K. K. and Toh, K. C. (2013), Effective block diagonal preconditioners for Biot's consolidation equations in piled-raft foundations. *International Journal for Numerical and Analytical Methods in Geomechanics*, 37(8), 871–892.

Clancy, P. and Randolph, M. F. (1993), An approximate analysis procedure for piled raft foundations. *International Journal for Numerical and Analytical Methods in Geomechanics*, 17(12), 849–869.

Cooke, R. W. (1986), Piled raft foundations on stiff clays: A contribution to design philosophy. *Géotechnique*, 36(2), 169–203.

Cunha, R., Poulos, H. and Small, J. (2001), Investigation of design alternatives for a piled raft case history. *Journal of Geotechnical and Geoenvironmental Engineering*, 127(8), 635–641.

DIN 1045:2005 (2005), Subsoil: Verification of the safety of soil engineering and foundations. German National Standard, Beuth Verlag, Berlin.

DIN EN 1997-1:2004 (2004), Eurocode 7 Geotechnical design—Part 1: General Rules. German National Standard, Beuth Verlag, Berlin.

Eslami, M. M., Aminikhah, A. and Ahmadi, M. M. (2011), A comparative study on pile group and piled raft foundations (PRF) behavior under seismic loading. *Computational Methods of Civil Engineering*, 2(2), 185–199.

Frank, R. (2010), Some aspects of research and practice for foundations designs in France, 11th Šuklje day, 17 September.

Hooper, J. A. (1972), Observations on the behavior of a piled-raft foundation on London Clay. *Proceedings Institution of Civil Engineers*, 55(2), 855–877.

Horikoshi, K., Masumoto, T., Hashizume, Y., Watanabe, T. and Fukuyama, H. (2003), Performance of piled raft foundations subjected to static horizontal loads. *International Journal of Physical Modelling in Geotechnics*, 2, 37–50.

Horikoshi, K. and Randolph, M. F. (1998), A contribution to optimal design of piled rafts. *Géotechnique*, 48(3), 301–317.

Katzenbach, R. (2005), Optimised design of high-rise building foundations in settlement-sensitive soils. *Proceedings of International Geotechnical Conference of Soil-Structure Interaction*, 26–28 May, St. Petersburg, Russia.

Katzenbach, R. and Reul, O. (1997), Design and performance of piled rafts. *Proceedings of 14th ICSMGE*, Hamburg, Germany, 4, 2253–2256.

Katzenbach, R., Bachmann, G., Boled- Mekasha, G. and Ramm, H. (2005), Combined pile-raft foundations (CPRF): An appropriate solution for the foundations of high rise buildings. *Slovak Journal of Civil Engineering*, 3, 19–29.

Katzenbach, R., Clauss, F., Ramm, H., Waberseck, T. and Choudhury, D. (2009), Combined pile-raft foundations and energy piles: Recent trends in research and practice. Proceedings of International Conference on Deep Foundations: CPRF and Energy Piles, 15 May, Frankfurt am Main, Germany, Darmstadt Geotechnics No. 18, Ed. R. Katzenbach, 3–20.

Katzenbach, R., Leppla, S., Ramm, H., Seip, M. and Kuttig, H. (2013), Design and construction of deep foundation systems and retaining structures in urban areas in difficult soil and groundwater conditions. *Procedia Engineering*, 57, 540–548.

Katzenbach, R., Ramm, H. and Choudhury, D. (2012), Combined pile-raft foundations: A sustainable foundations concept. *Proceedings of 9th International Conference on Testing and Design Methods for Deep Foundations*, 18–20 September, Kanazawa, Japan, p. 10.

Kitiyodom, P. and Matsumoto, T. (2003), A simplified analysis method for piled raft foundations in non-homogeneous soils. *International Journal for Numerical and Analytical Methods in Geomechanics*, 27(2), 85–109.

Kitiyodom, P., Matsumoto, T. and Kawaguchi, K. (2005), A simplified analysis method for piled raft foundations subjected to ground movements induced by tunneling. *International Journal for Numerical and Analytical Methods in Geomechanics*, 29(15), 1485–1507.

Leung, C. F. and Radhakrishnan, R. (1985), The behavior of pile-raft foundations in weak rock. *Proceedings of 11th ICSMFE*, San Francisco, CA, 3, 1429–1432.

Maharaj, D. K. and Gandhi, S. R. (2004), Non-linear finite element analysis of piled-raft foundations. *Proceedings of the ICE: Geotechnical Engineering*, 157(3), 107–113.

Mandolini, A., Laora, R. D. and Mascarucci, Y. (2013), Rational design of piled raft. *Procedia Engineering*, 57, 45–52.

Narasimhan, S. L. and Kameswara Rao, N. S. V. (1982), Finite element analysis of piled circular footings. *Proceedings of 4th ICNMG*, Edmonton, Canada, 2, 843–852.

O'Brien, A. S., Burland, J.-B. and Chapman, T. (2012), Rafts and piled rafts, Chapter 56. *ICE Manual of Geotechnical Engineering*, 853–886.

O'Neill, M. W. (1996), Case histories of pile-supported rafts. Report by ISSMFE Technical Committee No.18.

Peck, R. B. (1969), Advantages and limitations of the observational method in applied soil mechanics. *Géotechnique*, 19(2), 171–187.

Poulos, H. G. (1994), An approximate numerical analysis of pile-raft interaction. *International Journal for Numerical and Analytical Methods in Geomechanics*, 18(2), 73–92.

Poulos, H. G. (2001), Piled raft foundations: Design and applications. *Géotechnique*, 51(2), 95–113.

Poulos, H. G., Small, J. C., Ta, L. D., Sinha, J. and Chen, L. (1997), Comparison of some methods for analysis of piled rafts. *Proceedings of 14th ICSMGE*, Hamburg, Germany, 2 1119–1124.

Prakoso, W. and Kulhawy, F. (2001), Contributions to piled raft foundation design. *Journal of Geotechnical and Geoenvironmental Engineering*, 127(1), 17–24.

Randolph, M. F. (1983), Design of piled raft foundations. CUED/D: Soils TR 143, Cambridge University, UK.

Randolph, M. F. (1994), Design methods for pile groups and piled rafts. *Proceedings of 13th ISMGE*, New Delhi, India, 5, 61–82.

Randolph, M. F. and Clancy, P. (1993), Efficient design of piled rafts. *Proceedings of Deep Foundations on Bored and Auger Piles*, Ghent, Belgium, 119–130.

Ranganatham, B. V. and Kaniraj, S. R. (1978), Settlement of model pile foundations in sand. *Indian Geotechnical Journal*, 8(1), 1–26.

Reul, O. (2000), In-situ-Messungen und numerische Studien zum Tragverhalten der Kombinierten Pfahl-Plattengründung. Mitteilungen des Institutes und der Versuchsanstalt für Geotechnik der Technischen Universität Darmstadt, Germany, Heft 53.

Reul, O. and Randolph, M. F. (2004), Design strategies for piled rafts subjected to nonuniform vertical loading. *Journal of Geotechnical and Geoenvironmental Engineering*, 130(1), 1–13.

Reul, O. and Randolph, M. F. (2009), Optimised design of combined pile-raft foundations. *Darmstadt Geotechnics*, 18, Germany, 149–169.

Sales, M. M., Small, J. C. and Poulos, H. G. (2010), Compensated piled rafts in clayey souls: Behavior, measurements and predictions. *Canadian Geotechnical Journal*, 47, 327–345.

Small, J. C. and Liu, H. L. S. (2008), Time-settlement behavior of piled raft foundations using infinite elements. *Computers and Geotechnics*, 35(2), 187–195.

Yamashita, K., Kakurai, M. and Yamada, T. (1994), Investigation of a pile raft foundation on stiff clay. *Proceedings of 13th ICSMGE*, New Delhi, India, 2, 543–546.

Yamashita, K., Hamada, J., Onimaru, S., and Higashino, M. (2012), Seismic behavior of piled raft with ground improvement supporting a base-isolated building on soft ground in Tokyo. *Soils and Foundations*, 52, 1000–1015.

Index

Printed in the United States
by Baker & Taylor Publisher Services

Printed in the United States
by Baker & Taylor Publisher Services